MATLAB Codes for Finite Element Analysis

SOLID MECHANICS AND ITS APPLICATIONS
Volume 157

Series Editor: G.M.L. GLADWELL
Department of Civil Engineering
University of Waterloo
Waterloo, Ontario, Canada N2L 3GI

Aims and Scope of the Series

The fundamental questions arising in mechanics are: *Why?, How?,* and *How much?*
The aim of this series is to provide lucid accounts written by authoritative researchers
giving vision and insight in answering these questions on the subject of mechanics as it
relates to solids.

The scope of the series covers the entire spectrum of solid mechanics. Thus it includes
the foundation of mechanics; variational formulations; computational mechanics;
statics, kinematics and dynamics of rigid and elastic bodies: vibrations of solids and
structures; dynamical systems and chaos; the theories of elasticity, plasticity and
viscoelasticity; composite materials; rods, beams, shells and membranes; structural
control and stability; soils, rocks and geomechanics; fracture; tribology; experimental
mechanics; biomechanics and machine design.

The median level of presentation is the first year graduate student. Some texts are
monographs defining the current state of the field; others are accessible to final year
undergraduates; but essentially the emphasis is on readability and clarity.

For other titles published in this series, go to
www.springer.com/series/6557

MATLAB Codes for Finite Element Analysis

Solids and Structures

A.J.M. Ferreira

Universidade do Porto
Portugal

 Springer

A.J.M. Ferreira
Universidade do Porto
Fac. Engenharia
Rua Dr. Roberto Frias
4200-465 Porto
Portugal
ferreira@fe.up.pt

Additional material to this book can be downloaded from http://extras.springer.com.

ISBN 978-94-007-8955-5   ISBN 978-1-4020-9200-8 (eBook)

Printed on acid-free paper

9 8 7 6 5 4 3 2 1

springer.com

Preface

This book intend to supply readers with some MATLAB codes for finite element analysis of solids and structures.

After a short introduction to MATLAB, the book illustrates the finite element implementation of some problems by simple scripts and functions.

The following problems are discussed:

- Discrete systems, such as springs and bars
- Beams and frames in bending in 2D and 3D
- Plane stress problems
- Plates in bending
- Free vibration of Timoshenko beams and Mindlin plates, including laminated composites
- Buckling of Timoshenko beams and Mindlin plates

The book does not intends to give a deep insight into the finite element details, just the basic equations so that the user can modify the codes. The book was prepared for undergraduate science and engineering students, although it may be useful for graduate students.

The MATLAB codes of this book are included in the disk. Readers are welcomed to use them freely.

The author does not guarantee that the codes are error-free, although a major effort was taken to verify all of them. Users should use MATLAB 7.0 or greater when running these codes.

Any suggestions or corrections are welcomed by an email to ferreira@fe.up.pt.

Porto, Portugal, *António Ferreira*
2008

Contents

Chapter 1
Short introduction to MATLAB

1.1 Introduction

MATLAB is a commercial software and a trademark of The MathWorks, Inc., USA. It is an integrated programming system, including graphical interfaces and a large number of specialized toolboxes. MATLAB is getting increasingly popular in all fields of science and engineering.

This chapter will provide some basic notions needed for the understanding of the remainder of the book. A deeper study of MATLAB can be obtained from many MATLAB books and the very useful help of MATLAB.

1.2 Matrices

Matrices are the fundamental object of MATLAB and are particularly important in this book. Matrices can be created in MATLAB in many ways, the simplest one obtained by the commands

```
>> A=[1 2 3;4 5 6;7 8 9]
A =
     1     2     3
     4     5     6
     7     8     9
```

Note the semi-colon at the end of each matrix line. We can also generate matrices by pre-defined functions, such as random matrices

```
>> rand(3)
ans =
    0.8147    0.9134    0.2785
    0.9058    0.6324    0.5469
    0.1270    0.0975    0.9575
```

A.J.M. Ferreira, *MATLAB Codes for Finite Element Analysis:*
Solids and Structures, Solid Mechanics and Its Applications 157,
© Springer Science+Business Media B.V. 2009

Rectangular matrices can be obtained by specification of the number of rows and columns, as in

```
>> rand(2,3)
ans =
    0.9649    0.9706    0.4854
    0.1576    0.9572    0.8003
```

1.3 Operating with matrices

We can add, subtract, multiply, and transpose matrices. For example, we can obtain a matrix $C = A + B$, by the following commands

```
>> a=rand(4)
a =
    0.2769    0.6948    0.4387    0.1869
    0.0462    0.3171    0.3816    0.4898
    0.0971    0.9502    0.7655    0.4456
    0.8235    0.0344    0.7952    0.6463
>> b=rand(4)
b =
    0.7094    0.6551    0.9597    0.7513
    0.7547    0.1626    0.3404    0.2551
    0.2760    0.1190    0.5853    0.5060
    0.6797    0.4984    0.2238    0.6991
>> c=a+b
c =
    0.9863    1.3499    1.3985    0.9381
    0.8009    0.4797    0.7219    0.7449
    0.3732    1.0692    1.3508    0.9515
    1.5032    0.5328    1.0190    1.3454
```

The matrices can be multiplied, for example $E = A * D$, as shown in the following example

```
>> d=rand(4,1)
d =
    0.8909
    0.9593
    0.5472
    0.1386
>> e=a*d
e =
    1.1792
    0.6220
```

        1.4787
        1.2914

The transpose of a matrix is given by the apostrophe, as

```
>> a=rand(3,2)
a =
    0.1493    0.2543
    0.2575    0.8143
    0.8407    0.2435
>> a'
ans =
    0.1493    0.2575    0.8407
    0.2543    0.8143    0.2435
```

1.4 Statements

Statements are operators, functions and variables, always producing a matrix which can be used later. Some examples of statements:

```
>> a=3
a =
    3

>> b=a*3
b =
    9

>> eye(3)
ans =
    1    0    0
    0    1    0
    0    0    1
```

If one wants to cancel the echo of the input, a semi-colon at the end of the statement suffices.

Important to mention that MATLAB is case-sensitive, variables a and A being different objects.

We can erase variables from the workspace by using clear, or clear all. A given object can be erased, such as clear A.

1.5 Matrix functions

Some useful matrix functions are given in table 1.1

Table 1.1 Some useful functions for matrices

eye	Identity matrix
zeros	A matrix of zeros
ones	A matrix of ones
diag	Creates or extract diagonals
rand	Random matrix

Some examples of such functions are given in the following commands (here we build matrices by blocks)

```
>> [eye(3),diag(eye(3)),rand(3)]
ans =
    1.0000        0        0   1.0000   0.9293   0.2511   0.3517
         0   1.0000        0   1.0000   0.3500   0.6160   0.8308
         0        0   1.0000   1.0000   0.1966   0.4733   0.5853
```

Another example of matrices built from blocks:

```
>> A=rand(3)
A =
    0.5497   0.7572   0.5678
    0.9172   0.7537   0.0759
    0.2858   0.3804   0.0540
>> B = [A, zeros(3,2); zeros(2,3), ones(2)]
B =
    0.5497   0.7572   0.5678        0        0
    0.9172   0.7537   0.0759        0        0
    0.2858   0.3804   0.0540        0        0
         0        0        0   1.0000   1.0000
         0        0        0   1.0000   1.0000
```

1.6 Conditionals, if and switch

Often a function needs to branch based on runtime conditions. MATLAB offers structures for this similar to those in most languages. Here is an example illustrating most of the features of if.

```
x=-1
if x==0
    disp('Bad input!')
elseif max(x) > 0
    y = x+1;
else
    y = x^2;
end
```

If there are many options, it may better to use switch instead. For instance:

```
switch units
    case 'length'
        disp('meters')
    case 'volume'
        disp('cubic meters')
    case 'time'
        disp('hours')
    otherwise
        disp('not interested')
end
```

1.7 Loops: for and while

Many programs require iteration, or repetitive execution of a block of statements. Again, MATLAB is similar to other languages here. This code for calculating the first 10 Fibonacci numbers illustrates the most common type of for/end loop:

```
>> f=[1 2]
f =
     1     2
>> for i=3:10;f(i)=f(i-1)+f(i-2);end;
>> f
f =
     1     2     3     5     8    13    21    34    55    89
```

It is sometimes necessary to repeat statements based on a condition rather than a fixed number of times. This is done with while.

```
>> x=10;while x > 1; x = x/2,end
x =
     5
x =
    2.5000
x =
    1.2500
x =
    0.6250
```

Other examples of for/end loops:

```
>> x = []; for i = 1:4, x=[x,i^2], end
```

```
x =
    1
x =
    1    4
x =
    1    4    9
x =
    1    4    9   16
```

and in inverse form

```
>> x = []; for i = 4:-1:1, x=[x,i^2], end
x =
   16
x =
   16    9
x =
   16    9    4
x =
   16    9    4    1
```

Note the initial values of x = [] and the possibility of decreasing cycles.

1.8 Relations

Relations in MATLAB are shown in table 1.2.

Note the difference between '=' and logical equal '=='. The logical operators are given in table 1.3. The result if either 0 or 1, as in

```
>> 3<5,3>5,3==5
```

Table 1.2 Some relation operators

<	Less than
>	Greater than
<=	Less or equal than
>=	Greater or equal than
==	Equal to
~=	Not equal

Table 1.3 Logical operators

&	and
|	or
~	not

```
ans =
    1
ans =
    0
ans =
    0
```

The same is obtained for matrices, as in

```
>> a = rand(5), b = triu(a), a == b
a =
    0.1419    0.6557    0.7577    0.7060    0.8235
    0.4218    0.0357    0.7431    0.0318    0.6948
    0.9157    0.8491    0.3922    0.2769    0.3171
    0.7922    0.9340    0.6555    0.0462    0.9502
    0.9595    0.6787    0.1712    0.0971    0.0344
b =
    0.1419    0.6557    0.7577    0.7060    0.8235
         0    0.0357    0.7431    0.0318    0.6948
         0         0    0.3922    0.2769    0.3171
         0         0         0    0.0462    0.9502
         0         0         0         0    0.0344
ans =
    1    1    1    1    1
    0    1    1    1    1
    0    0    1    1    1
    0    0    0    1    1
    0    0    0    0    1
```

1.9 Scalar functions

Some MATLAB functions are applied to scalars only. Some of those functions are listed in table 1.4. Note that such functions can be applied to all elements of a vector or matrix, as in

```
>> a=rand(3,4)
a =
    0.4387    0.7952    0.4456    0.7547
```

Table 1.4 Scalar functions

sin	asin	exp	abs	round
cos	acos	log	sqrt	floor
tan	atan	rem	sign	ceil

```
     0.3816      0.1869      0.6463      0.2760
     0.7655      0.4898      0.7094      0.6797
>> b=sin(a)
b =
     0.4248      0.7140      0.4310      0.6851
     0.3724      0.1858      0.6022      0.2725
     0.6929      0.4704      0.6514      0.6286
>> c=sqrt(b)
c =
     0.6518      0.8450      0.6565      0.8277
     0.6102      0.4310      0.7760      0.5220
     0.8324      0.6859      0.8071      0.7928
```

1.10 Vector functions

Some MATLAB functions operate on vectors only, such as those illustrated in
table 1.5.

Consider for example vector X=1:10. The sum, mean and maximum values are
evaluated as

```
>> x=1:10
x =
     1     2     3     4     5     6     7     8     9    10
>> sum(x)
ans =
    55
>> mean(x)
ans =
    5.5000
>> max(x)
ans =
    10
```

Table 1.5 Vector functions

max	sum	median	any
min	prod	mean	all

1.11 Matrix functions

Some important matrix functions are listed in table 1.6.

In some cases such functions may use more than one output argument, as in

```
>> A=rand(3)
A =
    0.8147    0.9134    0.2785
    0.9058    0.6324    0.5469
    0.1270    0.0975    0.9575
>> y=eig(A)
y =
   -0.1879
    1.7527
    0.8399
```

where we wish to obtain the eigenvalues only, or in

```
>> [V,D]=eig(A)
V =
    0.6752   -0.7134   -0.5420
   -0.7375   -0.6727   -0.2587
   -0.0120   -0.1964    0.7996
D =
   -0.1879         0         0
         0    1.7527         0
         0         0    0.8399
```

where we obtain the eigenvectors and the eigenvalues of matrix A.

Table 1.6 Matrix functions

eig	Eigenvalues and eigenvectors
chol	Choleski factorization
inv	Inverse
lu	LU decomposition
qr	QR factorization
schur	Schur decomposition
poly	Characteristic polynomial
det	Determinant
size	Size of a matrix
norm	1-norm, 2-norm, F-norm, ∞-norm
cond	Conditioning number of 2-norm
rank	Rank of a matrix

1.12 Submatrix

In MATLAB it is possible to manipulate matrices in order to make code more compact or more efficient. For example, using the colon we can generate vectors, as in

```
>> x=1:8
x =
     1     2     3     4     5     6     7     8
```

or using increments

```
>> x=1.2:0.5:3.7
x =
    1.2000    1.7000    2.2000    2.7000    3.2000    3.7000
```

This sort of vectorization programming is quite efficient, no for/end cycles are used. This efficiency can be seen in the generation of a table of sines,

```
>> x=0:pi/2:2*pi
x =
         0    1.5708    3.1416    4.7124    6.2832
>> b=sin(x)
b =
         0    1.0000    0.0000   -1.0000   -0.0000
>> [x' b']
ans =
         0         0
    1.5708    1.0000
    3.1416    0.0000
    4.7124   -1.0000
    6.2832   -0.0000
```

The colon can also be used to access one or more elements from a matrix, where each dimension is given a single index or vector of indices. A block is then extracted from the matrix, as illustrated next.

```
>> a=rand(3,4)
a =
    0.6551    0.4984    0.5853    0.2551
    0.1626    0.9597    0.2238    0.5060
    0.1190    0.3404    0.7513    0.6991
>> a(2,3)
ans =
    0.2238
```

```
>> a(1:2,2:3)
ans =
    0.4984      0.5853
    0.9597      0.2238
>> a(1,end)
ans =
    0.2551
>> a(1,:)
ans =
    0.6551      0.4984      0.5853      0.2551
>> a(:,3)
ans =
    0.5853
    0.2238
    0.7513
```

It is interesting to note that arrays are stored linearly in memory, from the first dimension, second, and so on. So we can in fact access vectors by a single index, as show below.

```
>> a=[1 2 3;4 5 6; 9 8 7]
a =
    1      2      3
    4      5      6
    9      8      7
>> a(3)
ans =
    9
>> a(7)
ans =
    3
>> a([1 2 3 4])
ans =
    1      4      9      2
>> a(:)
ans =
    1
    4
    9
    2
    5
    8
    3
    6
    7
```

Subscript referencing can also be used in both sides.

```
>> a
a =
      1     2     3
      4     5     6
      9     8     7
>> b
b =
      1     2     3
      4     5     6
>> b(1,:)=a(1,:)
b =
      1     2     3
      4     5     6
>> b(1,:)=a(2,:)
b =
      4     5     6
      4     5     6
>> b(:,2)=[]
b =
      4     6
      4     6
>> a(3,:)=0
a =
      1     2     3
      4     5     6
      0     0     0
>> b(3,1)=20
b =
      4     6
      4     6
     20     0
```

As you noted in the last example, we can insert one element in matrix B, and MATLAB automatically resizes the matrix.

1.13 Logical indexing

Logical indexing arise from logical relations, resulting in a logical array, with elements 0 or 1.

```
>> a
a =
      1     2     3
      4     5     6
      0     0     0
```

```
>> a>2
ans =
     0     0     1
     1     1     1
     0     0     0
```

Then we can use such array as a mask to modify the original matrix, as shown next.

```
>> a(ans)=20
a =
     1     2    20
    20    20    20
     0     0     0
```

This will be very useful in finite element calculations, particularly when imposing boundary conditions.

1.14 M-files, scripts and functions

A M-file is a plain text file with MATLAB commands, saved with extension .m. The M-files can be scripts of functions. By using the editor of MATLAB we can insert comments or statements and then save or compile the m-file. Note that the percent sign % represents a comment. No statement after this sign will be executed. Comments are quite useful for documenting the file.

M-files are useful when the number of statements is large, or when you want to execute it at a later stage, or frequently, or even to run it in background.

A simple example of a **script** is given below.

```
% program 1
% programmer: Antonio ferreira
% date: 2008.05.30
% purpose : show how M-files are built

% data: a - matrix of numbers; b: matrix with sines of a


a=rand(3,4);
b=sin(a);
```

Functions act like subroutines in fortran where a particular set of tasks is performed. A typical function is given below, where in the first line we should name the function and give the input parameters (m,n,p) in parenthesis and the output parameters (a,b,c) in square parenthesis.

```
function  [a,b,c] = antonio(m,n,p)
```

```
a = hilb(m);
b= magic(n);
c= eye(m,p);
```

We then call this function as

```
>> [a,b,c]=antonio(2,3,4)
```

producing

```
>> [a,b,c]=antonio(2,3,4)
a =
    1.0000    0.5000
    0.5000    0.3333
b =
    8    1    6
    3    5    7
    4    9    2
c =
    1    0    0    0
    0    1    0    0
```

It is possible to use only some output parameters.

```
>> [a,b]=antonio(2,3,4)
a =
    1.0000    0.5000
    0.5000    0.3333
b =
    8    1    6
    3    5    7
    4    9    2
```

1.15 Graphics

MATLAB allows you to produce graphics in a simple way, either 2D or 3D plots.

1.15.1 2D plots

Using the command `plot` we can produce simple 2D plots in a **figure**, using two vectors with x and y coordinates. A simple example

```
x = -4:.01:4;  y = sin(x);  plot(x,y)
```

producing the plot of figure 1.1.

Fig. 1.1 2D plot of a
sinus

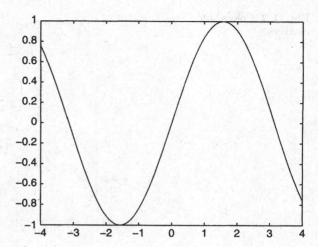

Table 1.7 Some graphics commands

Title	Title
xlabel	x-axis legend
ylabel	y-axis legend
Axis([x_{min},x_{max},y_{min},y_{max}])	Sets limits to axis
Axis auto	Automatic limits
Axis square	Same scale for both axis
Axis equal	Same scale for both axis
Axis off	Removes scale
Axis on	Scales again

We can insert a title, legends, modify axes etc., as shown in table 1.7.

By using `hold on` we can produce several plots in the same figure. We can also
modify colors of curves or points, as in

```
>> x=0:.01:2*pi; y1=sin(x); y2=sin(2*x); y3=sin(4*x);
>> plot(x,y1,'--',x,y2,':',x,y3,'+')
```

producing the plot of figure 1.2.

1.15.2 3D plots

As for 2D plots, we can produce 3D plots with `plot3` using x, y, and z vectors.
For example

```
t=.01:.01:20*pi; x=cos(t); y=sin(t); z=t.^3; plot3(x,y,z)
```

produces the plot illustrated in figure 1.3.

The next statements produce the graphic illustrated in figure 1.4.

Fig. 1.2 Colors and
markers

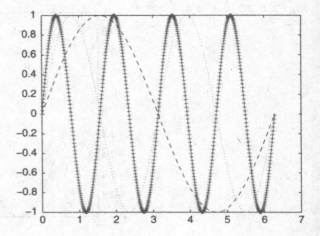

Fig. 1.3 3D plot

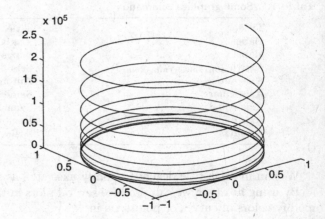

```
>> [xx,yy]=meshgrid(x,x);
>> z=exp(-xx.^2-yy.^2);
>> surf(xx,yy,z,gradient(z))
```

1.16 Linear algebra

In our finite element calculations we typically need to solve systems of equations,
or obtain the eigenvalues of a matrix. MATLAB has a large number of functions for
linear algebra. Only the most relevant for finite element analysis are here presented.

Fig. 1.4 Another 3D plot

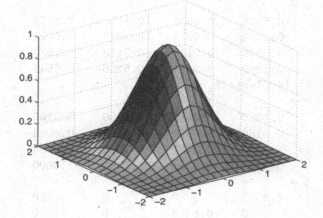

Consider a linear system $AX = B$, where

```
>> a=rand(3)
a =
    0.8909    0.1386    0.8407
    0.9593    0.1493    0.2543
    0.5472    0.2575    0.8143
>> b=rand(3,1)
b =
    0.2435
    0.9293
    0.3500
```

The solution vector X can be easily evaluated by using the backslash command,

```
>> x=a\b
x =
    0.7837
    2.9335
   -1.0246
```

Consider two matrices (for example a stiffness matrix and a mass matrix), for which we wish to calculate the generalized eigenproblem.

```
>> a=rand(4)
a =
    0.1966    0.3517    0.9172    0.3804
    0.2511    0.8308    0.2858    0.5678
    0.6160    0.5853    0.7572    0.0759
    0.4733    0.5497    0.7537    0.0540
>> b=rand(4)
```

```
b =
    0.5308    0.5688    0.1622    0.1656
    0.7792    0.4694    0.7943    0.6020
    0.9340    0.0119    0.3112    0.2630
    0.1299    0.3371    0.5285    0.6541
>> [v,d]=eig(a,b)
v =
    0.1886   -0.0955    1.0000   -0.9100
    0.0180    1.0000   -0.5159   -0.4044
   -1.0000   -0.2492   -0.2340    0.0394
    0.9522   -0.8833    0.6731   -1.0000
d =
   -4.8305         0         0         0
         0   -0.6993         0         0
         0         0    0.1822         0
         0         0         0    0.7628
```

The MATLAB function eig can be applied to the generalized eigenproblem, producing matrix V, each column containing an eigenvector, and matrix D, containing the eigenvalues at its diagonal. If the matrices are the stiffness and the mass matrices then the eigenvectors will be the modes of vibration and the eigenvalues will be the square roots of the natural frequencies of the system.

Chapter 2
Discrete systems

2.1 Introduction

The finite element method is nowadays the most used computational tool, in science and engineering applications. The finite element method had its origin around 1950, with reference works of Courant [1], Argyris [2] and Clough [3].

Many finite element books are available, such as the books by Reddy [4], Onate [5], Zienkiewicz [6], Hughes [7], Hinton [8], just to name a few. Some recent books deal with the finite element analysis with MATLAB codes [9,10]. The programming approach in these books is quite different from the one presented in this book.

In this chapter some basic concepts are illustrated by solving discrete systems built from springs and bars.

2.2 Springs and bars

Consider a bar (or spring) element with two nodes, two degrees of freedom, corresponding to two axial displacements $u_1^{(e)}, u_2^{(e)},$[1] as illustrated in figure 2.1. We suppose an element of length L, constant cross-section with area A, and modulus of elasticity E. The element supports axial forces only.

The deformation in the bar is obtained as

$$\epsilon = \frac{u_2 - u_1}{L^{(e)}} \tag{2.1}$$

while the stress in the bar is given by the Hooke's law as

$$\sigma = E^{(e)}\epsilon = E^{(e)}\frac{u_2 - u_1}{L^{(e)}} \tag{2.2}$$

---

[1] The superscript$^{(e)}$ refers to a generic finite element.

A.J.M. Ferreira, *MATLAB Codes for Finite Element Analysis:*
Solids and Structures, Solid Mechanics and Its Applications 157,
© Springer Science+Business Media B.V. 2009

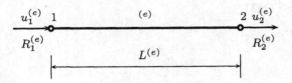

Fig. 2.1 Spring or bar finite element with two nodes

The axial resultant force is obtained by integration of stresses across the thickness direction as

$$N = A^{(e)}\sigma = (EA)^{(e)}\frac{u_2 - u_1}{L^{(e)}} \tag{2.3}$$

Taking into account the static equilibrium of the axial forces $R_1^{(e)}$ and $R_2^{(e)}$, as

$$R_2^{(e)} = -R_1^{(e)} = N = \left(\frac{EA}{L}\right)^{(e)} (u_2^{(e)} - u_1^{(e)}) \tag{2.4}$$

we can write the equations in the form (taking $k^{(e)} = \frac{EA}{L}$)

$$\mathbf{q}^{(e)} = \left\{ \begin{matrix} R_1^{(e)} \\ R_2^{(e)} \end{matrix} \right\} = k^{(e)} \begin{bmatrix} 1 & -1 \\ -1 & 1 \end{bmatrix} \left\{ \begin{matrix} u_1^{(e)} \\ u_2^{(e)} \end{matrix} \right\} = \mathbf{K}^{(e)}\mathbf{a}^{(e)} \tag{2.5}$$

where $\mathbf{K}^{(e)}$ is the stiffness matrix of the bar (spring) element, $\mathbf{a}^{(e)}$ is the displacement vector, and $\mathbf{q}^{(e)}$ represents the vector of nodal forces. If the element undergoes the action of distributed forces, it is necessary to transform those forces into nodal forces, by

$$\mathbf{q}^{(e)} = k^{(e)} \begin{bmatrix} 1 & -1 \\ -1 & 1 \end{bmatrix} \left\{ \begin{matrix} u_1^{(e)} \\ u_2^{(e)} \end{matrix} \right\} - \frac{(bl)^{(e)}}{2} \left\{ \begin{matrix} 1 \\ 1 \end{matrix} \right\} = \mathbf{K}^{(e)}\mathbf{a}^{(e)} - \mathbf{f}^{(e)} \tag{2.6}$$

with $\mathbf{f}^{(e)}$ being the vector of nodal forces equivalent to distributed forces $\mathbf{b}$.

2.3 Equilibrium at nodes

In (2.6) we show the equilibrium relation for one element, but we also need to obtain the equations of equilibrium for the structure. Therefore, we need to assemble the contribution of all elements so that a global system of equations can be obtained. To do that we recall that **in each node the sum of all forces arising from various adjacent elements equals the applied load at that node**.

We then obtain

$$\sum_{e=1}^{n_e} R^{(e)} = R_j^{(e)} \tag{2.7}$$

where n_e represents the number of elements in the structure, producing a global system of equations in the form

$$\begin{bmatrix} K_{11} & K_{12} & \dots & K_{1n} \\ K_{21} & K_{22} & \dots & K_{2n} \\ \vdots & \vdots & & \vdots \\ K_{n1} & K_{n2} & \dots & K_{nn} \end{bmatrix} \begin{Bmatrix} u_1 \\ u_2 \\ \vdots \\ u_n \end{Bmatrix} = \begin{Bmatrix} f_1 \\ f_2 \\ \vdots \\ f_n \end{Bmatrix}$$

or in a more compact form

$$\mathbf{Ka} = \mathbf{f} \tag{2.8}$$

Here $\mathbf{K}$ represents the system (or structure) stiffness matrix, $\mathbf{a}$ is the system displacement vector, and $\mathbf{f}$ represents the system force vector.

2.4 Some basic steps

In any finite element problem, some calculation steps are typical:

- Define a set of elements connected at nodes
- For each element, compute stiffness matrix $\mathbf{K}^{(e)}$, and force vector $\mathbf{f}^{(e)}$
- Assemble the contribution of all elements into the global system $\mathbf{Ka} = \mathbf{f}$
- Modify the global system by imposing essential (displacements) boundary conditions
- Solve the global system and obtain the global displacements $\mathbf{a}$
- For each element, evaluate the strains and stresses (post-processing)

2.5 First problem and first MATLAB code

To illustrate some of the basic concepts, and introduce the first MATLAB code, we consider a problem, illustrated in figure 2.2 where the central bar is defined as rigid. Our problem has three finite elements and four nodes. Three nodes are clamped, being the boundary conditions defined as $u_1 = u_3 = u_4 = 0$. In order to solve this problem, we set $k = 1$ for all springs and the external applied load at node 2 to be $P = 10$.

We can write, for each element in turn, the (local) equilibrium equation
Spring 1:

$$\begin{Bmatrix} R_1^{(1)} \\ R_2^{(1)} \end{Bmatrix} = k^{(1)} \begin{bmatrix} 1 & -1 \\ -1 & 1 \end{bmatrix} \begin{Bmatrix} u_1^{(1)} \\ u_2^{(1)} \end{Bmatrix}$$

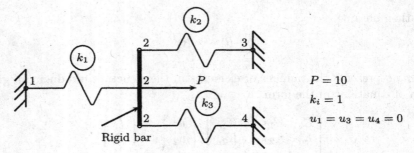

Fig. 2.2 Problem 1: a spring problem

Spring 2:

$$\left\{ \begin{array}{c} R_1^{(2)} \\ R_2^{(2)} \end{array} \right\} = k^{(2)} \begin{bmatrix} 1 & -1 \\ -1 & 1 \end{bmatrix} \left\{ \begin{array}{c} u_1^{(2)} \\ u_2^{(2)} \end{array} \right\}$$

Spring 3:

$$\left\{ \begin{array}{c} R_1^{(3)} \\ R_2^{(3)} \end{array} \right\} = k^{(3)} \begin{bmatrix} 1 & -1 \\ -1 & 1 \end{bmatrix} \left\{ \begin{array}{c} u_1^{(3)} \\ u_2^{(3)} \end{array} \right\}$$

We then consider the compatibility conditions to relate local (element) and global (structure) displacements as

$$u_1^{(1)} = u_1; \ u_2^{(1)} = u_2; \ u_1^{(2)} = u_2; \ u_2^{(2)} = u_3; \ u_1^{(3)} = u_2; \ u_2^{(3)} = u_4 \tag{2.9}$$

By expressing equilibrium of forces at nodes 1 to 4, we can write

$$\text{Node 1:} \sum_{e=1}^{3} R^{(e)} = F_1 \Leftrightarrow R_1^{(1)} = F_1 \tag{2.10}$$

$$\text{Node 2:} \sum_{e=1}^{3} R^{(e)} = P \Leftrightarrow R_2^{(1)} + R_1^{(2)} + R_1^{(3)} = P \tag{2.11}$$

$$\text{Node 3:} \sum_{e=1}^{3} R^{(e)} = F_3 \Leftrightarrow R_2^{(3)} = F_3 \tag{2.12}$$

$$\text{Node 4:} \sum_{e=1}^{3} R^{(e)} = F_4 \Leftrightarrow R_2^{(4)} = F_4 \tag{2.13}$$

and then obtain the static global equilibrium equations in the form

$$
\begin{bmatrix}
k_1 & -k_1 & 0 & 0 \\
-k_1 & k_1 + k_2 + k_3 & -k_2 & -k_3 \\
0 & -k_2 & k_2 & 0 \\
0 & -k_3 & 0 & k_3
\end{bmatrix}
\begin{Bmatrix}
u_1 \\ u_2 \\ u_3 \\ u_4
\end{Bmatrix}
=
\begin{Bmatrix}
F_1 \\ P \\ F_3 \\ F_4
\end{Bmatrix}
\tag{2.14}
$$

Taking into account the boundary conditions $u_1 = u_3 = u_4 = 0$, we may write

$$
\begin{bmatrix}
k_1 & -k_1 & 0 & 0 \\
-k_1 & k_1 + k_2 + k_3 & -k_2 & -k_3 \\
0 & -k_2 & k_2 & 0 \\
0 & -k_3 & 0 & k_3
\end{bmatrix}
\begin{Bmatrix}
0 \\ u_2 \\ 0 \\ 0
\end{Bmatrix}
=
\begin{Bmatrix}
F_1 \\ P \\ F_3 \\ F_4
\end{Bmatrix}
\tag{2.15}
$$

At this stage, we can compute the reactions F_1, F_3, F_4, only after the computation of the global displacements. We can remove lines and columns of the system, corresponding to $u_1 = u_3 = u_4 = 0$, and reduce the global system to one equation

$$
(k_1 + k_2 + k_3)u_2 = P
$$

The reactions can then be obtained by

$$
-k_1 u_2 = F_1; \quad -k_2 u_2 = F_3; \quad -k_3 u_2 = F_4
$$

Note that the stiffness matrix was obtained by "summing" the contributions of each element at the correct lines and columns corresponding to each element degrees of freedom. For instance, the degrees of freedom of element 1 are 1 and 2, and the 2×2 stiffness matrix of this element is placed at the corresponding lines and columns of the global stiffness matrix.

$$
K^{(1)} =
\begin{bmatrix}
k_1 & -k_1 & 0 & 0 \\
-k_1 & k_1 & 0 & 0 \\
0 & 0 & 0 & 0 \\
0 & 0 & 0 & 0
\end{bmatrix}
\tag{2.16}
$$

For element 3, the (global) degrees of freedom are 2 and 4 and the 2×2 stiffness matrix of this element is placed at the corresponding lines and columns of the global stiffness matrix.

$$
K^{(3)} =
\begin{bmatrix}
0 & 0 & 0 & 0 \\
0 & k_3 & 0 & -k_3 \\
0 & 0 & 0 & 0 \\
0 & -k_3 & 0 & k_3
\end{bmatrix}
\tag{2.17}
$$

A first MATLAB code problem1.m is introduced to solve the problem illustrated in figure 2.2. Many of the concepts used later on more complex elements are already given in this code. We set $k = 1$ for all elements and $P = 10$.

```
%..............................................................

% MATLAB codes for Finite Element Analysis
% problem1.m
% antonio ferreira 2008

% clear memory
clear all

% elementNodes: connections at elements
elementNodes=[1 2;2 3;2 4];

% numberElements: number of Elements
numberElements=size(elementNodes,1);

% numberNodes: number of nodes
numberNodes=4;

% for structure:
    % displacements: displacement vector
    % force : force vector
    % stiffness: stiffness matrix
displacements=zeros(numberNodes,1);
force=zeros(numberNodes,1);
stiffness=zeros(numberNodes);

% applied load at node 2
force(2)=10.0;

% computation of the system stiffness matrix
for e=1:numberElements;
  % elementDof: element degrees of freedom (Dof)
  elementDof=elementNodes(e,:) ;
  stiffness(elementDof,elementDof)=...
      stiffness(elementDof,elementDof)+[1 -1;-1 1];
end

% boundary conditions and solution
% prescribed dofs
prescribedDof=[1;3;4];
% free Dof : activeDof
activeDof=setdiff([1:numberNodes]',[prescribedDof]);
```

```
% solution
displacements=stiffness(activeDof,activeDof)\force(activeDof);

% positioning all displacements
displacements1=zeros(numberNodes,1);
displacements1(activeDof)=displacements;

% output displacements/reactions
outputDisplacementsReactions(displacements1,stiffness,...
    numberNodes,prescribedDof)
```

We discuss some of the programming steps. The workspace is deleted by

`clear all`

In matrix `elementNodes` we define the connections (left and right nodes) at each element,

`elementNodes=[1 2;2 3;2 4];`

In the first line of this matrix we place 1 and 2 corresponding to nodes 1 and 2, and proceed to the other lines in a similar way. By using the MATLAB function `size`, that returns the number of lines and columns of a rectangular matrix, we can detect the number of elements by inspecting the number of lines of matrix `elementNodes`.

```
% numberElements: number of Elements
numberElements=size(elementNodes,1);
```

Note that in this problem, the number of nodes is 4,

```
% numberNodes: number of nodes
numberNodes=4;
```

In this problem, the number of nodes is the same as the number of degrees of freedom (which is not the case in many other examples). Because the stiffness matrix is the result of an assembly process, involving summing of contributions, it is important to initialize it. Also, it is a good programming practice to do so in MATLAB, in order to increase the speed of `for` loops.

Using MATLAB function `zeros` we initialize the global displacement vector `displacement`, the global force vector `force` and the global stiffness matrix `stiffness`, respectively.

```
% for structure:
    % displacements: displacement vector
    % force : force vector
    % stiffness: stiffness matrix
```

```
displacements=zeros(numberNodes,1);
force=zeros(numberNodes,1);
stiffness=zeros(numberNodes);
```

We now place the applied force at the corresponding degree of freedom:

```
% applied load at node 2
force(2)=10.0;
```

We compute now the stiffness matrix for each element in turn and then assemble it in the global stiffness matrix.

```
% calculation of the system stiffness matrix
for e=1:numberElements;
  % elementDof: element degrees of freedom (Dof)
  elementDof=elementNodes(e,:) ;
  stiffness(elementDof,elementDof)=...
      stiffness(elementDof,elementDof)+[1 -1;-1 1];
end
```

In the first line of the cycle, we inspect the degrees of freedom at each element, in a vector `elementDof`. For example, for element 1, `elementDof` =[1,2], for element 2, `elementDof` =[2 3] and so on.

```
  % elementDof: element degrees of freedom (Dof)
  elementDof=elementNodes(e,:) ;
```

Next we state that the stiffness matrix for each element is constant and then we perform the assembly process by "spreading" this 2×2 matrix at the corresponding lines and columns defined by `elementDof` ,

```
  stiffness(elementDof,elementDof)=...
      stiffness(elementDof,elementDof)+[1 -1;-1 1];
```

The line `stiffness(elementDof,elementDof)+[1 -1;-1 1];` of the code can be interpreted as

```
stiffness([1 2],[1 2])= stiffness([1 2],[1 2])+[1 -1;-1 1];
```

for element 1,

```
stiffness([2 3],[2 3])= stiffness([2 3],[2 3])+[1 -1;-1 1];
```

for element 2, and

```
stiffness([2 4],[2 4])= stiffness([2 4],[2 4])+[1 -1;-1 1];
```

for element 3. This sort of coding allows a quick and compact assembly.

This global system of equations cannot be solved at this stage. We need to impose essential boundary conditions before solving the system $\mathbf{Ka} = \mathbf{f}$. The lines and columns of the prescribed degrees of freedom, as well as the lines of the force vector will be eliminated at this stage.

First we define vector **prescribedDof**, corresponding to the prescribed degrees of freedom. Then we define a vector containing all **activeDof** degrees of freedom, by setting up the difference between all degrees of freedom and the prescribed ones. The MATLAB function **setdiff** allows this operation.

```
% boundary conditions and solution
% prescribed dofs
prescribedDof=[1;3;4];
% free Dof : activeDof
activeDof=setdiff([1:numberNodes]',[prescribedDof]);
% solution
displacements=stiffness(activeDof,activeDof)\force(activeDof);
```

Note that the solution is performed with the active lines and columns only, by using a **mask**.

```
displacements=stiffness(activeDof,activeDof)\force(activeDof);
```

Because we are in fact calculating the solution for the active degrees of freedom only, we can place this solution in a vector **displacements1** that contains also the prescribed (zero) values.

```
% positioning all displacements
displacements1=zeros(numberNodes,1);
displacements1(activeDof)=displacements;
```

The vector **displacements1** is then a four-position vector with all displacements. We then call function outputDisplacementsReactions.m, to output displacements and reactions, as

```
%..........................................................

function outputDisplacementsReactions...
    (displacements,stiffness,GDof,prescribedDof)

% output of displacements and reactions in
% tabular form

% GDof: total number of degrees of freedom of
% the problem

% displacements
disp('Displacements')
%displacements=displacements1;
jj=1:GDof; format
[jj' displacements]
```

```
% reactions
F=stiffness*displacements;
reactions=F(prescribedDof);
disp('reactions')
[prescribedDof reactions]
```

Reactions are computed by evaluating the total force vector as F = K.U. Because we only need reactions (forces at prescribed degrees of freedom), we then use

```
% reactions
F=stiffness*displacements;
reactions=F(prescribedDof);
```

When running this code we obtain detailed information on matrices or results, depending on the user needs, for example displacements and reactions:

```
Displacements

ans =

    1.0000          0
    2.0000     3.3333
    3.0000          0
    4.0000          0

reactions

ans =

    1.0000    -3.3333
    3.0000    -3.3333
    4.0000    -3.3333
```

2.6 New code using MATLAB structures

MATLAB structures provide a way to collect arrays of different types and sizes in a single array. The following codes show how this can be made for problem1.m. One of the most interesting features is that the argument passing to functions is simplified.

The new problem1Structure.m listing is as follows:

```
%.............................................................
% MATLAB codes for Finite Element Analysis
% problem1Structure.m
% antonio ferreira 2008

% clear memory
clear all

% p1 : structure
p1=struct();

% elementNodes: connections at elements
p1.elementNodes=[1 2;2 3;2 4];

% GDof: total degrees of freedom
p1.GDof=4;

% numberElements: number of Elements
p1.numberElements=size(p1.elementNodes,1);

% numberNodes: number of nodes
p1.numberNodes=4;

% for structure:
    % displacements: displacement vector
    % force : force vector
    % stiffness: stiffness matrix
p1.displacements=zeros(p1.GDof,1);
p1.force=zeros(p1.GDof,1);
p1.stiffness=zeros(p1.GDof);

% applied load at node 2
p1.force(2)=10.0;

% computation of the system stiffness matrix
for e=1:p1.numberElements;
  % elementDof: element degrees of freedom (Dof)
  elementDof=p1.elementNodes(e,:) ;
  p1.stiffness(elementDof,elementDof)=...
      p1.stiffness(elementDof,elementDof)+[1 -1;-1 1];
end
```

```
% boundary conditions and solution
% prescribed dofs
p1.prescribedDof=[1;3;4];

% solution
p1.displacements=solutionStructure(p1)

% output displacements/reactions
outputDisplacementsReactionsStructure(p1)
```

Note that p is now a structure that collects most of the problem data:

p1 =

```
    elementNodes:  [3x2 double]
            GDof:  4
   numberElements: 3
      numberNodes:  4
    displacements:  [4x1 double]
            force:  [4x1 double]
        stiffness:  [4x4 double]
    prescribedDof:  [3x1 double]
```

Information on a particular data field can be accessed as follows:

```
>> p1.prescribedDof

ans =

    1
    3
    4
```

or

```
>> p1.displacements

ans =

         0
    3.3333
         0
         0
```

This code calls solutionStructure.m which computes displacements by eliminating lines and columns of prescribed degrees of freedom. Notice that the argument passing is made by just providing the structure name (p1). This is simpler than passing all relevant matrices.

```
%..............................................................................

function displacements=solutionStructure(p)
% function to find solution in terms of global displacements
activeDof=setdiff([1:p.GDof]', [p.prescribedDof]);
U=p.stiffness(activeDof,activeDof)\p.force(activeDof);
displacements=zeros(p.GDof,1);
displacements(activeDof)=U;
```

It also calls the function outputDisplacementsReactionsStructure.m which outputs displacements and reactions.

```
%..............................................................................

function outputDisplacementsReactionsStructure(p)

% output of displacements and reactions in
% tabular form

% GDof: total number of degrees of freedom of
% the problem

% displacements
disp('Displacements')
jj=1:p.GDof; format
[jj' p.displacements]

% reactions
F=p.stiffness*p.displacements;
reactions=F(p.prescribedDof);
disp('reactions')
[p.prescribedDof reactions]
```

Chapter 3
Analysis of bars

3.1 A bar element

Consider the two-node bar finite element shown in figure 3.1, with constant cross-section (area A) and length $L = 2a$. The bar element can undergo only axial stresses σ_x, which are uniform in every cross-section.

The work stored as strain energy dU is obtained as

$$dU = \frac{1}{2}\sigma_x \epsilon_x A dx \qquad (3.1)$$

The total strain energy is given by

$$U = \frac{1}{2}\int_{-a}^{a} \sigma_x \epsilon_x A dx \qquad (3.2)$$

By assuming a linear elastic behaviour of the bar material, we can write

$$\sigma_x = E\epsilon_x \qquad (3.3)$$

where E is the modulus of elasticity. Therefore the strain energy can be expressed as

$$U = \frac{1}{2}\int_{-a}^{a} EA\epsilon_x^2 dx \qquad (3.4)$$

Strains ϵ_x can be related with the axial displacements u as

$$\epsilon_x = \frac{du}{dx} \qquad (3.5)$$

By substituting (3.5) into (3.4) we then obtain

$$U = \frac{1}{2}\int_{-a}^{a} EA\left(\frac{du}{dx}\right)^2 dx \qquad (3.6)$$

A.J.M. Ferreira, *MATLAB Codes for Finite Element Analysis:*
Solids and Structures, Solid Mechanics and Its Applications 157,
© Springer Science+Business Media B.V. 2009

Fig. 3.1 A bar element in its local coordinate system

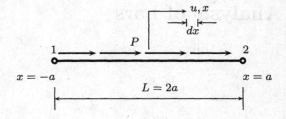

Fig. 3.2 A two-node bar element

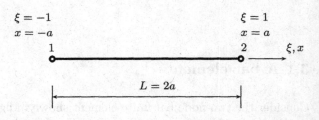

If we consider p as the applied forces by unit length, the virtual external work at each element is

$$\delta W = \int_{-a}^{a} p\delta u dx \tag{3.7}$$

Let's consider now a two-noded finite element, as illustrated in figure 3.2. The axial displacements can be interpolated as

$$u = N_1(\xi)u_1 + N_2(\xi)u_2 \tag{3.8}$$

where the shape functions are defined as

$$N_1(\xi) = \frac{1}{2}(1-\xi); \quad N_2(\xi) = \frac{1}{2}(1+\xi) \tag{3.9}$$

in the natural coordinate system $\xi \in [-1, +1]$. The interpolation (3.8) can be defined in matrix form as

$$\mathbf{u} = \begin{bmatrix} N_1 & N_2 \end{bmatrix} \begin{bmatrix} u_1 \\ u_2 \end{bmatrix} = \mathbf{N}\mathbf{u}^e \tag{3.10}$$

The element strain energy is now expressed as

$$U = \frac{1}{2}\int_{-a}^{a} EA\left(\frac{du}{dx}\right)^2 dx = \frac{1}{2}\int_{-1}^{1} \frac{EA}{a^2}\left(\frac{du}{d\xi}\right)^2 a d\xi = \frac{1}{2}\mathbf{u}^{eT}\int_{-1}^{1} \frac{EA}{a}\mathbf{N'}^T\mathbf{N'}d\xi\mathbf{u}^e \tag{3.11}$$

where $\mathbf{N'} = \dfrac{du}{d\xi}$, and

$$U = \frac{1}{2}\mathbf{u}^{eT}\mathbf{K}^e\mathbf{u}^e \tag{3.12}$$

The element stiffness matrix, $\mathbf{K}^e$, is given by

$$\mathbf{K}^e = \frac{EA}{a} \int_{-1}^{1} \mathbf{N}'^T \mathbf{N}' d\xi \tag{3.13}$$

The integral is evaluated in the natural system, after a transformation of coordinates $x = a\xi$, including the evaluation of the jacobian determinant, $|J| = \frac{dx}{d\xi} = a$.

In this element the derivatives of the shape functions are

$$\frac{dN_1}{d\xi} = -\frac{1}{2}; \quad \frac{dN_2}{d\xi} = \frac{1}{2} \tag{3.14}$$

In this case, the stiffness matrix can be given in explicit form as

$$\mathbf{K}^e = \frac{EA}{a} \int_{-1}^{1} \begin{bmatrix} -\frac{1}{2} \\ \frac{1}{2} \end{bmatrix} \begin{bmatrix} -\frac{1}{2} & \frac{1}{2} \end{bmatrix} d\xi = \frac{EA}{2a} \begin{bmatrix} 1 & -1 \\ -1 & 1 \end{bmatrix} \tag{3.15}$$

By using $L = 2a$ we obtain the same stiffness matrix as in the direct method presented in the previous chapter.

The virtual work done by the external forces is defined as

$$\delta W^e = \int_{-a}^{a} p\delta u\, dx = \int_{-1}^{1} p\delta u a\, d\xi = \delta \mathbf{u}^{eT} a \int_{-1}^{1} p\mathbf{N}^T d\xi \tag{3.16}$$

or

$$\delta W^e = \delta \mathbf{u}^{eT} \mathbf{f}^e \tag{3.17}$$

where the vector of nodal forces that are equivalent to distributed forces is given by

$$\mathbf{f}^e = a \int_{-1}^{1} p\mathbf{N}^T d\xi = \frac{ap}{2} \int_{-1}^{1} \begin{bmatrix} 1-\xi \\ 1+\xi \end{bmatrix} d\xi = ap \begin{bmatrix} 1 \\ 1 \end{bmatrix} \tag{3.18}$$

For a system of bars, the contribution of each element must be assembled. For example in the bar of figure 3.3, we consider five nodes and four elements. In this case the structure vector of displacements is given by

$$\mathbf{u}^T = \begin{bmatrix} u_1 & u_2 & u_3 & u_4 & u_5 \end{bmatrix} \tag{3.19}$$

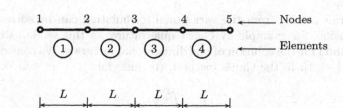

Fig. 3.3 Bar discretized into four elements

Summing the contribution of all elements, we obtain the strain energy and the energy done by the external forces as

$$U = \frac{1}{2}\mathbf{u}^T \sum_{e=1}^{4} \mathbf{K}^e \mathbf{u} = \frac{1}{2}\mathbf{u}^T \mathbf{K} \mathbf{u} \tag{3.20}$$

$$\delta W = \delta \mathbf{u}^T \sum_{e=1}^{4} \mathbf{f}^e = \delta \mathbf{u}^T \mathbf{f} \tag{3.21}$$

where $\mathbf{K}$ and $\mathbf{f}$ are the structure stiffness matrix and the force vector, respectively.

The stiffness matrix is then assembled as

$$\mathbf{K} = \frac{EA}{L}\left\{ \underbrace{\begin{bmatrix} 1 & -1 & 0 & 0 & 0 \\ -1 & 1 & 0 & 0 & 0 \\ 0 & 0 & 0 & 0 & 0 \\ 0 & 0 & 0 & 0 & 0 \\ 0 & 0 & 0 & 0 & 0 \end{bmatrix}}_{\text{element 1}} + \underbrace{\begin{bmatrix} 0 & 0 & 0 & 0 & 0 \\ 0 & 1 & -1 & 0 & 0 \\ 0 & -1 & 1 & 0 & 0 \\ 0 & 0 & 0 & 0 & 0 \\ 0 & 0 & 0 & 0 & 0 \end{bmatrix}}_{\text{element 2}} + ... \right\} = \frac{EA}{L}\begin{bmatrix} 1 & -1 & 0 & 0 & 0 \\ -1 & 2 & -1 & 0 & 0 \\ 0 & -1 & 2 & -1 & 0 \\ 0 & 0 & -1 & 2 & -1 \\ 0 & 0 & 0 & -1 & 1 \end{bmatrix} \tag{3.22}$$

whereas the vector of equivalent forces is given by

$$\mathbf{f} = ap \begin{bmatrix} 1 \\ 2 \\ 2 \\ 2 \\ 1 \end{bmatrix} \tag{3.23}$$

We then obtain a global system of equations

$$\mathbf{K}\mathbf{u} = \mathbf{f} \tag{3.24}$$

to be solved after the imposition of the boundary conditions as explained before.

3.2 Numerical integration

The integrals arising from the variational formulation can be solved by numerical integration, for example by Gauss quadrature. In this section we present the Gauss method for the solution of one dimensional integrals. We consider a function $f(x)$, $x \in [-1,1]$. In the Gauss method, the integral

$$I = \int_{-1}^{1} f(x)dx \tag{3.25}$$

Table 3.1 Coordinates and weights for Gauss integration

n	$\pm x_i$	W_i
1	0.0	2.0
2	0.5773502692	1.0
3	0.774596697	0.5555555556
	0.0	0.8888888889
4	0.86113663116	0.3478548451
	0.3399810436	0.6521451549

Fig. 3.4 One dimensional Gauss quadrature for two and one Gauss points

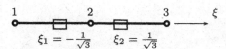

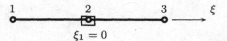

is replaced by a sum of p Gauss points, in which the function at those points is multiplied by some weights, as in

$$I = \int_{-1}^{1} f(x)dx = \sum_{i=1}^{p} f(x_i)W_i \qquad (3.26)$$

where W_i is the i-th point weight. In table 3.1 the coordinates and weights of the Gauss technique are presented. This technique is exact for a $2n - 1$ polynomial if we use at least n Gauss points. In figure 3.4 the Gauss points positions are illustrated.

3.3 An example of isoparametric bar

MATLAB code problem2.m solves the bar problem illustrated in figure 3.5, in which the modulus of elasticity is $E = 30e6$, and the area of the cross-section is $A = 1$.

The code problem2.m considers an isoparametric formulation.

```
%..........................................................................

% MATLAB codes for Finite Element Analysis
```

Fig. 3.5 Clamped bar subjected to
point load, problem2.m

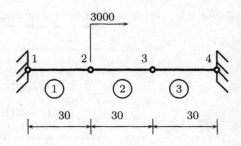

```
% problem2.m
% antonio ferreira 2008

% clear memory
clear all

% E; modulus of elasticity
% A: area of cross section
% L: length of bar
E  = 30e6;A=1;EA=E*A; L  = 90;

% generation of coordinates and connectivities
% numberElements: number of elements
numberElements=3;
% generation equal spaced coordinates
nodeCoordinates=linspace(0,L,numberElements+1);
xx=nodeCoordinates;
% numberNodes: number of nodes
numberNodes=size(nodeCoordinates,2);

% elementNodes: connections at elements
ii=1:numberElements;
elementNodes(:,1)=ii;
elementNodes(:,2)=ii+1;

% for structure:
    % displacements: displacement vector
    % force : force vector
    % stiffness: stiffness matrix
displacements=zeros(numberNodes,1);
force=zeros(numberNodes,1);
stiffness=zeros(numberNodes,numberNodes);
```

```
% applied load at node 2
force(2)=3000.0;

% computation of the system stiffness matrix
for e=1:numberElements;
  % elementDof: element degrees of freedom (Dof)
  elementDof=elementNodes(e,:) ;
  nn=length(elementDof);
  length_element=nodeCoordinates(elementDof(2))...
      -nodeCoordinates(elementDof(1));
  detJacobian=length_element/2;invJacobian=1/detJacobian;

  % central Gauss point (xi=0, weight W=2)
  [shape,naturalDerivatives]=shapeFunctionL2(0.0);
    Xderivatives=naturalDerivatives*invJacobian;

% B matrix
B=zeros(1,nn);  B(1:nn)  = Xderivatives(:);
  stiffness(elementDof,elementDof)=...
      stiffness(elementDof,elementDof)+B'*B*2*detJacobian*EA;
end

% boundary conditions and solution
% prescribed dofs
fixedDof=find(xx==min(nodeCoordinates(:)) ...
    | xx==max(nodeCoordinates(:)))';
prescribedDof=[fixedDof]
% free Dof : activeDof
activeDof=setdiff([1:numberNodes]',[prescribedDof]);

% solution
GDof=numberNodes;
displacements=solution(GDof,prescribedDof,stiffness,force);

% output displacements/reactions
outputDisplacementsReactions(displacements,stiffness,...
    numberNodes,prescribedDof)
```

The nodal coordinates are obtained by an equal-spaced division of the domain, using linspace.

```
% generation of coordinates and connectivities
% numberElements: number of elements
numberElements=3;
```

```
% generation equal spaced coordinates
nodeCoordinates=linspace(0,L,numberElements+1);
```

The connectivities are obtained by a vectorized cycle

```
% elementNodes: connections at elements
ii=1:numberElements;
elementNodes(:,1)=ii;
elementNodes(:,2)=ii+1;
```

We use a Gauss quadrature with one central point $\xi = 0$ and weight 2. The evaluation of the stiffness matrix involves the integral (3.15) by

```
stiffness(elementDof,elementDof)=...
    stiffness(elementDof,elementDof)+B'*B*2*detJacobian*EA;
```

where **B** is a matrix with the derivatives of the shape functions

```
% B matrix
B=zeros(1,nn);  B(1:nn)  = Xderivatives(:);
  stiffness(elementDof,elementDof)=...
    stiffness(elementDof,elementDof)+B'*B*2*detJacobian*EA;
```

The shape function and derivatives with respect to natural coordinates are computed in function shapeFunctionL2.m.

```
% .............................................................
    function [shape,naturalDerivatives]=shapeFunctionL2(xi)

    % shape function and derivatives for L2 elements
    % shape : Shape functions
    % naturalDerivatives: derivatives w.r.t. xi
    % xi: natural coordinates (-1 ... +1)

    shape=([1-xi,1+xi]/2)';

    naturalDerivatives=[-1;1]/2;

    end % end function shapeFunctionL2
```

The function (solution.m) will be used in the remaining of the book. This function computes the displacements of any FE system in the forthcoming problems.

```
%.............................................................

function displacements=solution(GDof,prescribedDof,stiffness,
    force)
% function to find solution in terms of global displacements
activeDof=setdiff([1:GDof]', ...
    [prescribedDof]);
U=stiffness(activeDof,activeDof)\force(activeDof);
displacements=zeros(GDof,1);
displacements(activeDof)=U;
```

3.4 Problem 2, using MATLAB struct

Another possible code using MATLAB structures would be problem2Structure.m,
as:

```
%.............................................................

% MATLAB codes for Finite Element Analysis
% problem2Structure.m
% antonio ferreira 2008

% clear memory
clear all

% p1 : structure
p1=struct();

% E; modulus of elasticity
% A: area of cross section
% L: length of bar
E  = 30e6;A=1;EA=E*A; L  = 90;

% generation of coordinates and connectivities
% numberElements: number of elements
p1.numberElements=3;
% generation equal spaced coordinates
p1.nodeCoordinates=linspace(0,L,p1.numberElements+1);
```

```
p1.xx=p1.nodeCoordinates;
% numberNodes: number of nodes
p1.numberNodes=size(p1.nodeCoordinates,2);

% elementNodes: connections at elements
ii=1:p1.numberElements;
p1.elementNodes(:,1)=ii;
p1.elementNodes(:,2)=ii+1;

% GDof: total degrees of freedom
p1.GDof=p1.numberNodes;

% % numberElements: number of Elements
% p1.numberElements=size(p1.elementNodes,1);
%
% % numberNodes: number of nodes
% p1.numberNodes=4;

% for structure:
    % displacements: displacement vector
    % force : force vector
    % stiffness: stiffness matrix
p1.displacements=zeros(p1.GDof,1);
p1.force=zeros(p1.GDof,1);
p1.stiffness=zeros(p1.GDof);

% applied load at node 2
p1.force(2)=3000.0;

% computation of the system stiffness matrix
for e=1:p1.numberElements;
  % elementDof: element degrees of freedom (Dof)
  elementDof=p1.elementNodes(e,:) ;
  nn=length(elementDof);
  length_element=p1.nodeCoordinates(elementDof(2))...
      -p1.nodeCoordinates(elementDof(1));
  detJacobian=length_element/2;invJacobian=1/detJacobian;

  % central Gauss point (xi=0, weight W=2)
  shapeL2=shapeFunctionL2Structure(0.0);
    Xderivatives=shapeL2.naturalDerivatives*invJacobian;

    % B matrix
```

```
   B=zeros(1,nn);  B(1:nn)  = Xderivatives(:);
   p1.stiffness(elementDof,elementDof)=...
       p1.stiffness(elementDof,elementDof)+B'*B*2*detJacobian*EA;
end                          .

% prescribed dofs
p1.prescribedDof=find(p1.xx==min(p1.nodeCoordinates(:)) ...
     | p1.xx==max(p1.nodeCoordinates(:)))';

% solution
p1.displacements=solutionStructure(p1)

% output displacements/reactions
outputDisplacementsReactionsStructure(p1)
```

The code needs some change in function shapeFunctionL2Structure.m, as follows:

```
%  ....................................................................
     function shapeL2=shapeFunctionL2Structure(xi)

     % shape function and derivatives for L2 elements
     % shape : Shape functions
     % naturalDerivatives: derivatives w.r.t. xi
     % xi: natural coordinates (-1 ... +1)

     shapeL2=struct()
     shapeL2.shape=([1-xi,1+xi]/2)';
     shapeL2.naturalDerivatives=[-1;1]/2;

     end % end function shapeFunctionL2
```

Results are placed in structure p1, as before.

```
>> p1

p1 =

    numberElements: 3
   nodeCoordinates: [0 30 60 90]
                xx: [0 30 60 90]
       numberNodes: 4
      elementNodes: [3x2 double]
```

```
        GDof:  4
displacements:  [4x1 double]
        force:  [4x1 double]
    stiffness:  [4x4 double]
prescribedDof:  [2x1 double]
```

We can obtain detailed information on p1, for example, displacements:

```
>> p1.displacements
```

```
ans =
```

```
         0
    0.0020
    0.0010
         0
```

3.5 Problem 3

Another problem involving bars and springs is illustrated in figure 3.6. The MAT-LAB code for this problem is problem3.m, using direct stiffness method.

```
%............................................................................

% MATLAB codes for Finite Element Analysis
% problem3.m
% ref: D. Logan, A first couse in the finite element method,
% third Edition, page 121, exercise P3-10
% direct stiffness method
% antonio ferreira 2008
```

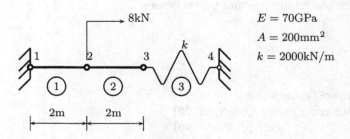

Fig. 3.6 Illustration of problem 3, problem3.m

```
% clear memory
clear all

% E; modulus of elasticity
% A: area of cross section
% L: length of bar
% k: spring stiffness
E=70000;A=200;k=2000;

% generation of coordinates and connectivities
% numberElements: number of elements
numberElements=3;
numberNodes=4;
elementNodes=[1 2; 2 3; 3 4];
nodeCoordinates=[0 2000 4000 4000];
xx=nodeCoordinates;

% for structure:
    % displacements: displacement vector
    % force : force vector
    % stiffness: stiffness matrix
displacements=zeros(numberNodes,1);
force=zeros(numberNodes,1);
stiffness=zeros(numberNodes,numberNodes);

% applied load at node 2
force(2)=8000;

% computation of the system stiffness matrix
for e=1:numberElements;
 % elementDof: element degrees of freedom (Dof)
 elementDof=elementNodes(e,:) ;
 L=nodeCoordinates(elementDof(2))-nodeCoordinates(elementDof(1));
  if e<3
      ea(e)=E*A/L;
  else
      ea(e)=k;
  end
  stiffness(elementDof,elementDof)=...
      stiffness(elementDof,elementDof)+ea(e)*[1 -1;-1 1];
end
% boundary conditions and solution
```

```
% prescribed dofs
prescribedDof=[1;4];
% free Dof : activeDof
activeDof=setdiff([1:numberNodes]',[prescribedDof]);

% solution
GDof=4;
displacements=solution(GDof,prescribedDof,stiffness,force);

% output displacements/reactions
outputDisplacementsReactions(displacements,stiffness,...
    numberNodes,prescribedDof)
```

The isoparametric version for the problem illustrated in figure 3.6 is given in problem3a.m.

```
%.................................................................

% MATLAB codes for Finite Element Analysis
% problem3a.m
% ref: D. Logan, A first couse in the finite element method,
% third Edition, page 121, exercise P3-10
% with isoparametric formulation
% antonio ferreira 2008

% clear memory
clear all

% E; modulus of elasticity
% A: area of cross section
% L: length of bar
E  = 70000;A=200;EA=E*A;k=2000;

% generation of coordinates and connectivities
numberElements=3;
numberNodes=4;
elementNodes=[1 2; 2 3; 3 4];
nodeCoordinates=[0 2000 4000 4000];
xx=nodeCoordinates;

% for structure:
```

```
    % displacements: displacement vector
    % force : force vector
    % stiffness: stiffness matrix
displacements=zeros(numberNodes,1);
force=zeros(numberNodes,1);
stiffness=zeros(numberNodes,numberNodes);

% applied load at node 2
force(2)=8000.0;

% computation of the system stiffness matrix
for e=1:numberElements;
  % elementDof: element degrees of freedom (Dof)
  elementDof=elementNodes(e,:) ;

  if e<3 % bar elements
  nn=length(elementDof);
  length_element=nodeCoordinates(elementDof(2))...
      -nodeCoordinates(elementDof(1));
  detJacobian=length_element/2;invJacobian=1/detJacobian;
% central Gauss point (xi=0, weight W=2)
  % central Gauss point (xi=0, weight W=2)
  [shape,naturalDerivatives]=shapeFunctionL2(0.0);
    Xderivatives=naturalDerivatives*invJacobian;

% B matrix
B=zeros(1,nn);  B(1:nn)  = Xderivatives(:);
  ea(e)=E*A;
  stiffness(elementDof,elementDof)=...
    stiffness(elementDof,elementDof)+B'*B*2*detJacobian*ea(e);
  else % spring element
      stiffness(elementDof,elementDof)=...
      stiffness(elementDof,elementDof)+k*[1 -1;-1 1];
  end
end

% boundary conditions and solution
prescribedDof=[1;4];
GDof=4;

% solution
displacements=solution(GDof,prescribedDof,stiffness,force);
```

```
% output displacements/reactions
outputDisplacementsReactions(displacements,stiffness,...
    numberNodes,prescribedDof)
```

For both codes, the solution is the same and matches the analytical solution presented in Logan [11]. The displacements at nodes 2 and 3 are 0.935 and 0.727 mm, respectively. The reactions at the supports 1 and 4 are −6.546 and −1.455 kN, respectively.

The code problem3Structure.m is equivalent to problem3.m, but using MATLAB structures.

```
%............................................................

% MATLAB codes for Finite Element Analysis
% problem3Structure.m
% antonio ferreira 2008

% clear memory
clear all

% p1 : structure
p1=struct();

% E; modulus of elasticity
% A: area of cross section
% L: length of bar
% k: spring stiffness
E=70000;A=200;k=2000;

% generation of coordinates and connectivities
% numberElements: number of elements
p1.numberElements=3;
p1.numberNodes=4;
p1.elementNodes=[1 2; 2 3; 3 4];
p1.nodeCoordinates=[0 2000 4000 4000];
p1.xx=p1.nodeCoordinates;

% GDof: total degrees of freedom
p1.GDof=p1.numberNodes;

% for structure:
```

```
    % displacements: displacement vector
    % force : force vector
    % stiffness: stiffness matrix
p1.displacements=zeros(p1.GDof,1);
p1.force=zeros(p1.GDof,1);
p1.stiffness=zeros(p1.GDof);

% applied load at node 2
p1.force(2)=8000.0;

% computation of the system stiffness matrix
for e=1:p1.numberElements;
  % elementDof: element degrees of freedom (Dof)
  elementDof=p1.elementNodes(e,:) ;
  L=p1.nodeCoordinates(elementDof(2))-...
      p1.nodeCoordinates(elementDof(1));
  if e<3
      ea(e)=E*A/L;
  else
      ea(e)=k;
  end
  p1.stiffness(elementDof,elementDof)=...
      p1.stiffness(elementDof,elementDof)+ea(e)*[1 -1;-1 1];
end

% prescribed dofs
p1.prescribedDof=[1;4];

% solution
p1.displacements=solutionStructure(p1)

% output displacements/reactions
outputDisplacementsReactionsStructure(p1)
```

Chapter 4
Analysis of 2D trusses

4.1 Introduction

This chapter deals with the static analysis of two dimensional trusses, which are basically bars oriented in two dimensional cartesian systems. A transformation of coordinate basis is necessary to translate the local element matrices (stiffness matrix, force vector) into the structural (global) coordinate system. Trusses support compressive and tensile forces only, as in bars. All forces are applied at the nodes. After the presentation of the element formulation, some examples are solved by MATLAB codes.

4.2 2D trusses

In figure 4.1 we consider a typical 2D truss in global $x - y$ plane. The local system of coordinates $x' - y'$ defines the local displacements u'_1, u'_2. The element possesses two degrees of freedom in the local setting,

$$\mathbf{u}'^T = [u'_1 \quad u'_2] \tag{4.1}$$

while in the global coordinate system, the element is defined by four degrees of freedom

$$\mathbf{u}^T = [u_1 \quad u_2 \quad u_3 \quad u_4] \tag{4.2}$$

The relation between both local and global displacements is given by

$$u'_1 = u_1 cos(\theta) + u_2 sin(\theta) \tag{4.3}$$

$$u'_2 = u_3 cos(\theta) + u_4 sin(\theta) \tag{4.4}$$

A.J.M. Ferreira, *MATLAB Codes for Finite Element Analysis:*
Solids and Structures, Solid Mechanics and Its Applications 157,
© Springer Science+Business Media B.V. 2009

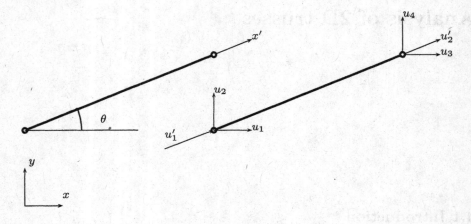

Fig. 4.1 2D truss element: local and global degrees of freedom

where θ is the angle between local axis x' and global axis x, or in matrix form as

$$\mathbf{u}' = \mathbf{L}\mathbf{u} \tag{4.5}$$

being matrix $\mathbf{L}$ defined as

$$\mathbf{L} = \begin{bmatrix} l & m & 0 & 0 \\ 0 & 0 & l & m \end{bmatrix} \tag{4.6}$$

The l, m elements of matrix L can be defined by the nodal coordinates as

$$l = \frac{x_2 - x_1}{L_e}; \quad m = \frac{y_2 - y_1}{L_e} \tag{4.7}$$

being L_e the length of the element,

$$L_e = \sqrt{(x_2 - x_1)^2 + (y_2 - y_1)^2} \tag{4.8}$$

4.3 Stiffness matrix

In the local coordinate system, the stiffness matrix of the 2D truss element is given by the bar stiffness, as before:

$$\mathbf{K}' = \frac{EA}{L_e} \begin{bmatrix} 1 & -1 \\ -1 & 1 \end{bmatrix} \tag{4.9}$$

In the local coordinate system, the strain energy of this element is given by

$$U_e = \frac{1}{2}\mathbf{u'}^T\mathbf{K'}\mathbf{u'} \tag{4.10}$$

Replacing $\mathbf{u'} = \mathbf{Lu}$ in (4.10) we obtain

$$U_e = \frac{1}{2}\mathbf{u}^T[\mathbf{L}^T\mathbf{K'}\mathbf{L}]\mathbf{u} \tag{4.11}$$

It is now possible to express the global stiffness matrix as

$$\mathbf{K} = \mathbf{L}^T\mathbf{K'}\mathbf{L} \tag{4.12}$$

or

$$\mathbf{K} = \frac{EA}{L_e}\begin{bmatrix} l^2 & lm & -l^2 & -lm \\ lm & m^2 & -lm & -m^2 \\ -l^2 & -lm & l^2 & lm \\ -lm & -m^2 & lm & m^2 \end{bmatrix} \tag{4.13}$$

4.4 Stresses at the element

In the local coordinate system, the stresses are defined as $\sigma = E\epsilon$. Taking into account the definition of strain in the bar, we obtain

$$\sigma = E\frac{u'_2 - u'_1}{L_e} = \frac{E}{L_e}[-1 \quad 1]\left\{\begin{array}{c} u'_1 \\ u'_2 \end{array}\right\} = \frac{E}{L_e}[-1 \quad 1]\mathbf{u'} \tag{4.14}$$

By transformation of local to global coordinates, we obtain stresses as function of the displacements as

$$\sigma = \frac{E}{L_e}[-1 \quad 1]\mathbf{Lu} = \frac{E}{L_e}[-l \quad -m \quad l \quad m]\mathbf{u} \tag{4.15}$$

4.5 First 2D truss problem

In a first 2D truss problem, illustrated in figure 4.2, we consider a downward point force (10,000) applied at node 1. The modulus of elasticity is $E = 30e6$ and all elements are supposed to have constant cross-section area $A = 2$. The supports are located in nodes 2 and 4. The numbering of degrees of freedom is illustrated in figure 4.3.

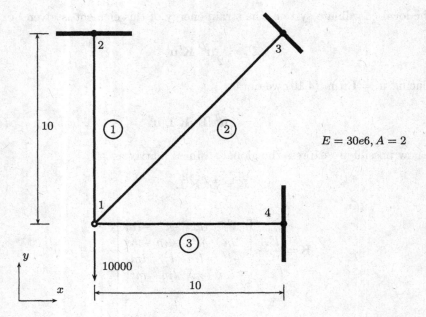

Fig. 4.2 First 2D truss problem, problem4.m

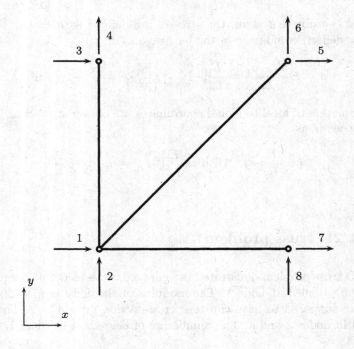

Fig. 4.3 First 2D truss problem: degrees of freedom

The code (problem4.m) listing is as:

```
%..............................................................

% MATLAB codes for Finite Element Analysis
% problem4.m
% antonio ferreira 2008

% clear memory
clear all

% E; modulus of elasticity
% A: area of cross section
% L: length of bar
E=30e6;    A=2;    EA=E*A;

% generation of coordinates and connectivities
numberElements=3;
numberNodes=4;
elementNodes=[1 2;1 3;1 4];
nodeCoordinates=[ 0 0;0 120;120 120;120 0];
xx=nodeCoordinates(:,1);
yy=nodeCoordinates(:,2);

% for structure:
    % displacements: displacement vector
    % force : force vector
    % stiffness: stiffness matrix

GDof=2*numberNodes; % GDof: total number of degrees of freedom
displacements=zeros(GDof,1);
force=zeros(GDof,1);

% applied load at node 2
force(2)=-10000.0;

% computation of the system stiffness matrix
  [stiffness]=...
    formStiffness2Dtruss(GDof,numberElements,...
    elementNodes,numberNodes,nodeCoordinates,xx,yy,EA);

% boundary conditions and solution
prescribedDof=[3:8]';
```

```
% solution
displacements=solution(GDof,prescribedDof,stiffness,force);

% drawing displacements
us=1:2:2*numberNodes-1;
vs=2:2:2*numberNodes;
figure
L=xx(2)-xx(1);
%L=node(2,1)-node(1,1);
XX=displacements(us);YY=displacements(vs);
dispNorm=max(sqrt(XX.^2+YY.^2));
scaleFact=15000*dispNorm;
clf
hold on
drawingMesh(nodeCoordinates+scaleFact*[XX YY],elementNodes,'L2',
    'k.-');
drawingMesh(nodeCoordinates,elementNodes,'L2','k.--');

% stresses at elements
stresses2Dtruss(numberElements,elementNodes,...
    xx,yy,displacements,E)

% output displacements/reactions
outputDisplacementsReactions(displacements,stiffness,...
    GDof,prescribedDof)
```

Note that this code calls some new functions. The first function
(formStiffness2Dtruss.m) computes the stiffness matrix of the 2D truss two-node
element.

```
function [stiffness]=...
    formStiffness2Dtruss(GDof,numberElements,...
    elementNodes,numberNodes,nodeCoordinates,xx,yy,EA);


stiffness=zeros(GDof);

% computation of the system stiffness matrix
for e=1:numberElements;
  % elementDof: element degrees of freedom (Dof)
  indice=elementNodes(e,:)    ;
```

```
    elementDof=[ indice(1)*2-1 indice(1)*2 indice(2)*2-1
        indice(2)*2] ;
    xa=xx(indice(2))-xx(indice(1));
    ya=yy(indice(2))-yy(indice(1));
    length_element=sqrt(xa*xa+ya*ya);
    C=xa/length_element;
    S=ya/length_element;
        k1=EA/length_element*...
            [C*C C*S -C*C -C*S; C*S S*S -C*S -S*S;
            -C*C -C*S C*C C*S;-C*S -S*S C*S S*S];
    stiffness(elementDof,elementDof)=...
            stiffness(elementDof,elementDof)+k1;
end
```

The function (stresses2Dtruss.m) computes stresses of the 2D truss elements.

```
function stresses2Dtruss(numberElements,elementNodes,...
    xx,yy,displacements,E)


% stresses at elements
for e=1:numberElements
    indice=elementNodes(e,:);
    elementDof=[ indice(1)*2-1 indice(1)*2 indice(2)*2-1
        indice(2)*2] ;
    xa=xx(indice(2))-xx(indice(1));
    ya=yy(indice(2))-yy(indice(1));
    length_element=sqrt(xa*xa+ya*ya);
    C=xa/length_element;
    S=ya/length_element;
    sigma(e)=E/length_element* ...
        [-C  -S C S]*displacements(elementDof);
end
disp('stresses')
sigma'
```

The code problem4.m is therefore easier to read by using functions that can also be used for other 2D truss problems.

Displacements, reactions and stresses are in full agreement with analytical results by Logan [11].

Displacements

ans =

```
    1.0000    0.0041
    2.0000   -0.0159
    3.0000         0
    4.0000         0
    5.0000         0
    6.0000         0
    7.0000         0
    8.0000         0
```

reactions

ans =

```
  1.0e+03 *

    0.0030         0
    0.0040    7.9289
    0.0050    2.0711
    0.0060    2.0711
```

stresses

ans =

```
  1.0e+03 *

    3.9645
    1.4645
   -1.0355
```

The deformation of the structure is illustrated in figure 4.4. We use a drawing routine drawingMesh for the purpose. This routine needs the input of nodal coordinates and elements connectivities and draws either undeformed and deformed meshes.

4.6 A second truss problem

The next problem is illustrated in figure 4.5. The degrees of freedom are illustrated in figure 4.6. The MATLAB code is problem5.m. The analytical solution of this problem is presented in [11]. The results of this code agree well with the analytical solution, although the analytical solution considered only half of the structure.

Fig. 4.4 Deformed shape
of 2D truss

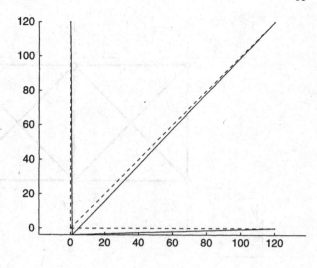

$$E = 70\text{GPa}, \, A = 3e^{-4}m^2$$

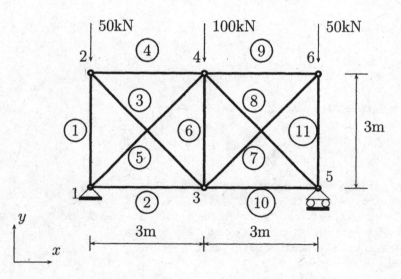

Fig. 4.5 A second truss problem, problem5.m

```
%...........................................

% MATLAB codes for Finite Element Analysis
% problem5.m
% antonio ferreira 2008
```

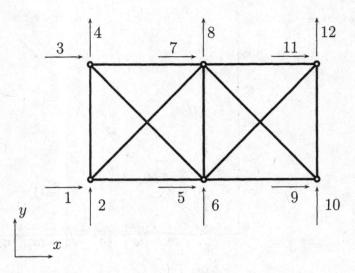

Fig. 4.6 A second truss problem: degrees of freedom

```
% clear memory
clear all

% E; modulus of elasticity
% A: area of cross section
% L: length of bar
E=70000;    A=300;    EA=E*A;

% generation of coordinates and connectivities
elementNodes=[ 1 2;1 3;2 3;2 4;1 4;3 4;3 6;4 5;4 6;3 5;5 6];
nodeCoordinates=[ 0 0;0 3000;3000 0;3000 3000;6000 0;6000 3000];
numberElements=size(elementNodes,1);
numberNodes=size(nodeCoordinates,1);
xx=nodeCoordinates(:,1);
yy=nodeCoordinates(:,2);

% for structure:
    % displacements: displacement vector
    % force : force vector
    % stiffness: stiffness matrix
GDof=2*numberNodes;
U=zeros(GDof,1);
force=zeros(GDof,1);
% applied load at node 2
```

```
force(4)=-50000;
force(8)=-100000;
force(12)=-50000;

% computation of the system stiffness matrix
 [stiffness]=...
     formStiffness2Dtruss(GDof,numberElements,...
     elementNodes,numberNodes,nodeCoordinates,xx,yy,EA);

% boundary conditions and solution
prescribedDof=[1 2 10]';

% solution
displacements=solution(GDof,prescribedDof,stiffness,force);
us=1:2:2*numberNodes-1;
vs=2:2:2*numberNodes;

% drawing displacements

figure
L=xx(2)-xx(1);
%L=node(2,1)-node(1,1);
XX=displacements(us);YY=displacements(vs);
dispNorm=max(sqrt(XX.^2+YY.^2));
scaleFact=2*dispNorm;
clf
hold on
drawingMesh(nodeCoordinates+scaleFact*[XX YY],...
     elementNodes,'L2','k.-');
drawingMesh(nodeCoordinates,elementNodes,'L2','k.--');

% output displacements/reactions
outputDisplacementsReactions(displacements,stiffness,...
    GDof,prescribedDof)

% stresses at elements
stresses2Dtruss(numberElements,elementNodes,...
    xx,yy,displacements,E)
```

Results are the following:

Displacements

ans =

```
     1.0000          0
     2.0000          0
     3.0000     7.1429
     4.0000    -9.0386
     5.0000     5.2471
     6.0000   -16.2965
     7.0000     5.2471
     8.0000   -20.0881
     9.0000    10.4942
    10.0000          0
    11.0000     3.3513
    12.0000    -9.0386
```

reactions

ans =

```
    1.0e+05 *

    0.0000     0.0000
    0.0000     1.0000
    0.0001     1.0000
```

stresses

ans =

```
 -210.9015
  122.4318
   62.5575
  -44.2349
 -173.1447
  -88.4697
   62.5575
 -173.1447
  -44.2349
  122.4318
 -210.9015
```

>>

Fig. 4.7 Deformed
shape, problem 5

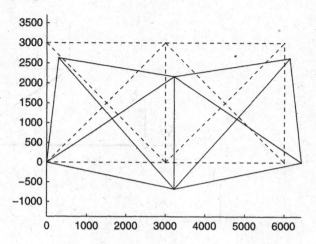

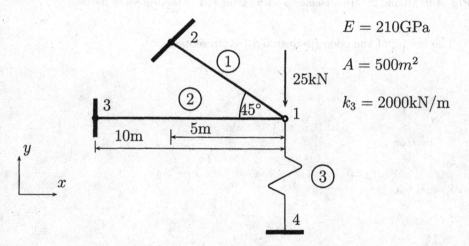

$E = 210\text{GPa}$

$A = 500m^2$

25kN

$k_3 = 2000\text{kN/m}$

Fig. 4.8 Mixing 2D truss elements with spring elements, problem6.m

The deformed shape of this problem is shown in figure 4.7.

4.7 An example of 2D truss with spring

In figure 4.8 we consider a structure that is built from two truss elements and one spring element. For the truss elements the modulus of elasticity is $E = 210$ GPa, and the cross-section area is $A = 5e^{-4}$ m^2. This problem is modeled with four points and three elements. Figure 4.9 illustrates the degrees of freedom according to our finite element discretization.

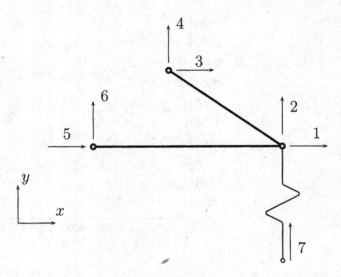

Fig. 4.9 Mixing 2D truss elements with spring elements: degrees of freedom

The listing of the code (**problem6.m**) is presented.

```
%.................................................

% MATLAB codes for Finite Element Analysis
% problem6.m
% ref: D. Logan, A first couse in the finite element method,
% third Edition, mixing trusses with springs
% antonio ferreira 2008

% clear memory
clear all

% E; modulus of elasticity
% A: area of cross section
% L: length of bar
E=210000;   A=500;     EA=E*A;

% generation of coordinates and connectivities
nodeCoordinates=[0 0;-5000*cos(pi/4) 5000*sin(pi/4); -10000 0];
elementNodes=[ 1 2;1 3;1 4];
numberElements=size(elementNodes,1);
numberNodes=size(nodeCoordinates,1)+1; % spring added
xx=nodeCoordinates(:,1);
yy=nodeCoordinates(:,2);
```

```
% for structure:
    % displacements: displacement vector
    % force : force vector
    % stiffness: stiffness matrix
GDof=2*numberNodes;
U=zeros(GDof,1);
force=zeros(GDof,1);
stiffness=zeros(GDof);

% applied load at node 2
force(2)=-25000;

% computation of the system stiffness matrix
for e=1:numberElements-1;
  % elementDof: element degrees of freedom (Dof)
  indice=elementNodes(e,:)    ;
  elementDof=[ indice(1)*2-1 indice(1)*2 indice(2)*2-1
    indice(2)*2] ;
  xa=xx(indice(2))-xx(indice(1));
  ya=yy(indice(2))-yy(indice(1));
  length_element=sqrt(xa*xa+ya*ya);
  C=xa/length_element;
  S=ya/length_element;
    k1=EA/length_element*...
        [C*C C*S -C*C -C*S; C*S S*S -C*S -S*S;
        -C*C -C*S C*C C*S;-C*S -S*S C*S S*S];
  stiffness(elementDof,elementDof)=...
      stiffness(elementDof,elementDof)+k1;
end
% spring stiffness in global Dof
stiffness([2 7],[2 7])= stiffness([2 7],[2 7])+2000*[1 -1;-1 1];

% boundary conditions and solution
prescribedDof=[3:8]';

% solution
displacements=solution(GDof,prescribedDof,stiffness,force);

% output displacements/reactions
outputDisplacementsReactions(displacements,stiffness,...
    GDof,prescribedDof)
```

```
% stresses at elements
for e=1:numberElements-1
  indice=elementNodes(e,:);
  elementDof=[ indice(1)*2-1 indice(1)*2 indice(2)*2-1
    indice(2)*2] ;
  xa=xx(indice(2))-xx(indice(1));
  ya=yy(indice(2))-yy(indice(1));
  length_element=sqrt(xa*xa+ya*ya);
  C=xa/length_element;
  S=ya/length_element;
  sigma(e)=E/length_element* ...
      [-C  -S C S]*displacements(elementDof);
end
disp('stresses')
sigma'
```

Note that the spring stiffness is added to global degrees of freedom 2 and 7, corresponding to vertical displacements at nodes 1 and 4.

Displacements, reactions and stresses are listed below. Displacements are exactly the same as the analytical solution [11]. Stresses in bars show that bar 1 is under tension and bar 2 is under compression.

```
Displacements

ans =

    1.0000    -1.7241
    2.0000    -3.4483
    3.0000         0
    4.0000         0
    5.0000         0
    6.0000         0
    7.0000         0
    8.0000         0

reactions

ans =

  1.0e+04 *

    0.0003    -1.8103
    0.0004     1.8103
    0.0005     1.8103
```

```
   0.0006           0
   0.0007      0.6897
   0.0008           0
```

stresses

ans =

```
   51.2043
  -36.2069
```

>>

Chapter 5
Trusses in 3D space

5.1 Basic formulation

We consider now trusses in 3D space. A typical two-noded 3D truss element is illustrated in figure 5.1. Each node has three global degrees of freedom.

The stiffness matrix in local coordinates is given by

$$K = (\frac{EA}{L}) \begin{bmatrix} C_x^2 & C_xC_y & C_xC_z & -C_x^2 & -C_xC_y & -C_xC_z \\ & C_y^2 & C_yC_z & -C_xC_y & -C_y^2 & -C_yC_z \\ & & C_z^2 & -C_xC_z & -C_yC_z & -C_z^2 \\ & & & C_x^2 & C_xC_y & C_xC_z \\ & & & & C_y^2 & C_yC_z \\ \text{symmetry} & & & & & C_z^2 \end{bmatrix}$$

where the cosines are obtained as

$$C_x = \frac{x_2 - x_1}{L}; C_y = \frac{y_2 - y_1}{L}; C_z = \frac{z_2 - z_1}{L}$$

We then perform a basis transformation in order to obtain the global stiffness matrix and the global vector of equivalent nodal forces, as we did for 2D trusses.

5.2 A 3D truss problem

We consider the 3D truss problem illustrated in figure 5.2. The MATLAB code (problem7.m) is used to evaluate displacements, reactions and forces at elements.

A.J.M. Ferreira, *MATLAB Codes for Finite Element Analysis:*
Solids and Structures, Solid Mechanics and Its Applications 157,
© Springer Science+Business Media B.V. 2009

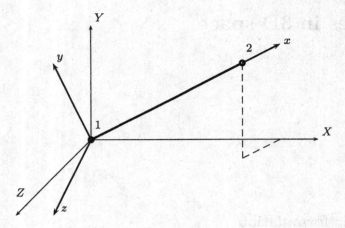

Fig. 5.1 Trusses in 3D coordinates: local and global coordinate sets

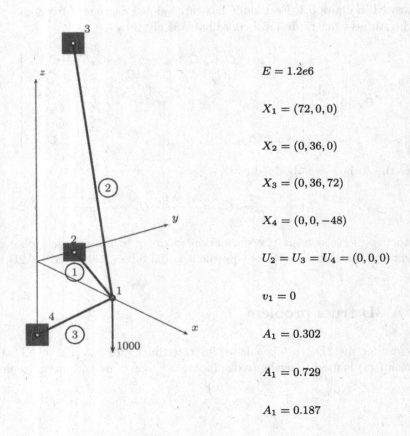

$E = 1.2e6$

$X_1 = (72, 0, 0)$

$X_2 = (0, 36, 0)$

$X_3 = (0, 36, 72)$

$X_4 = (0, 0, -48)$

$U_2 = U_3 = U_4 = (0, 0, 0)$

$v_1 = 0$

$A_1 = 0.302$

$A_1 = 0.729$

$A_1 = 0.187$

Fig. 5.2 A 3D truss problem: geometry, mesh, loads and boundary nodes, problem7.m

```
%..............................................................

% MATLAB codes for Finite Element Analysis
% problem7.m
% ref: D. Logan, A first couse in the finite element method,
% third Edition, A 3D truss example
% antonio ferreira 2008

% clear memory
clear all

% E; modulus of elasticity
% A: area of cross section
% L: length of bar
E=1.2e6;
A=[0.302;0.729;0.187]; % area for various sections

% generation of coordinates and connectivities
nodeCoordinates=[72 0 0; 0 36 0;  0 36 72; 0 0 -48];
elementNodes=[1 2;1 3;1 4];
numberElements=size(elementNodes,1);
numberNodes=size(nodeCoordinates,1);
xx=nodeCoordinates(:,1);
yy=nodeCoordinates(:,2);

% for structure:
    % displacements: displacement vector
    % force : force vector
    % stiffness: stiffness matrix
    % GDof: global number of degrees of freedom
GDof=3*numberNodes;
U=zeros(GDof,1);
force=zeros(GDof,1);

% applied load at node 2
force(3)=-1000;

% stiffness matrix
[stiffness]=...
    formStiffness3Dtruss(GDof,numberElements,...
    elementNodes,numberNodes,nodeCoordinates,E,A);
```

```
% boundary conditions and solution
prescribedDof=[2 4:12]';

% solution
displacements=solution(GDof,prescribedDof,stiffness,force);

% output displacements/reactions
outputDisplacementsReactions(displacements,stiffness,...
    GDof,prescribedDof)

% stresses at elements
stresses3Dtruss(numberElements,elementNodes,nodeCoordinates,...
    displacements,E)
```

The results are in excellent agreement with analytical solution in [11]:

Displacements

ans =

```
     1.0000    -0.0711
     2.0000         0
     3.0000    -0.2662
     4.0000         0
     5.0000         0
     6.0000         0
     7.0000         0
     8.0000         0
     9.0000         0
    10.0000         0
    11.0000         0
    12.0000         0
```

reactions

ans =

```
     2.0000  -223.1632
     4.0000   256.1226
     5.0000  -128.0613
     6.0000         0
     7.0000  -702.4491
     8.0000   351.2245
     9.0000   702.4491
```

```
10.0000   446.3264
11.0000          0
12.0000   297.5509
```

```
Stresses in elements
  1 -948.19142387
  2 1445.36842298
  3 -2868.54330060
```

5.3 A second 3D truss example

In figure 5.3 is illustrated a second example of a 3D truss.

```
%............................................................

% MATLAB codes for Finite Element Analysis
% problem8.m
% ref: D. Logan, A first couse in the finite element method,
% third Edition, A second 3D truss example
% antonio ferreira 2008
```

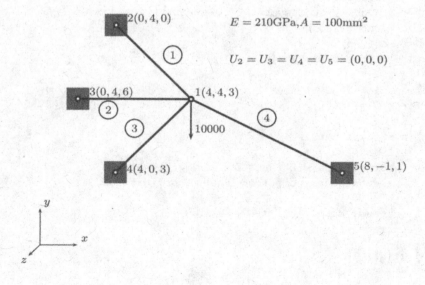

Fig. 5.3 Second 3D problem, problem8.m

```
% clear memory
clear all

% E; modulus of elasticity
% A: area of cross section
% L: length of bar
E=210000;
A=[100 100 100 100]; % area for various sections

% generation of coordinates and connectivities
nodeCoordinates=[4000 4000 3000;
                    0 4000 0;
                    0 4000 6000;
                 4000 0    3000;
                 8000 -1000 1000];
elementNodes=[1 2;1 3;1 4;1 5];
numberElements=size(elementNodes,1);
numberNodes=size(nodeCoordinates,1);
xx=nodeCoordinates(:,1);
yy=nodeCoordinates(:,2);

% for structure:
    % displacements: displacement vector
    % force : force vector
    % stiffness: stiffness matrix
    % GDof: global number of degrees of freedom
GDof=3*numberNodes;
U=zeros(GDof,1);
force=zeros(GDof,1);

% applied load at node 2
force(2)=-10000;

% stiffness matrix
[stiffness]=...
    formStiffness3Dtruss(GDof,numberElements,...
    elementNodes,numberNodes,nodeCoordinates,E,A);

% boundary conditions and solution
prescribedDof=[4:15]';

% solution
```

```
displacements=solution(GDof,prescribedDof,stiffness,force);

% output displacements/reactions
outputDisplacementsReactions(displacements,stiffness,...
    GDof,prescribedDof)

% stresses at elements
stresses3Dtruss(numberElements,elementNodes,nodeCoordinates,...
    displacements,E)
```

The results are in excellent agreement with analytical solution in [11]:

Displacements

ans =

```
     1.0000    -0.3024
     2.0000    -1.5177
     3.0000     0.2688
     4.0000          0
     5.0000          0
     6.0000          0
     7.0000          0
     8.0000          0
     9.0000          0
    10.0000          0
    11.0000          0
    12.0000          0
    13.0000          0
    14.0000          0
    15.0000          0
```

reactions

ans =

```
    1.0e+03 *

     0.0040     0.2709
     0.0050          0
     0.0060     0.2032
     0.0070     1.3546
     0.0080          0
     0.0090    -1.0160
```

```
   0.0100        0
   0.0110    7.9681
   0.0120        0
   0.0130   -1.6255
   0.0140    2.0319
   0.0150    0.8128


Stresses in elements
  1  -3.38652236
  2 -16.93261180
  3 -79.68086584
  4 -27.26097914
>>
>>
```

Both codes problem7.m and problem8.m call functions formStiffness3Dtruss.m for stiffness computation

```
function  [stiffness]=...
    formStiffness3Dtruss(GDof,numberElements,...
    elementNodes,numberNodes,nodeCoordinates,E,A);

stiffness=zeros(GDof);
% computation of the system stiffness matrix
for e=1:numberElements;
  % elementDof: element degrees of freedom (Dof)
  indice=elementNodes(e,:)   ;
  elementDof=[3*indice(1)-2 3*indice(1)-1 3*indice(1)...
          3*indice(2)-2 3*indice(2)-1 3*indice(2)] ;
  x1=nodeCoordinates(indice(1),1);
  y1=nodeCoordinates(indice(1),2);
  z1=nodeCoordinates(indice(1),3);
  x2=nodeCoordinates(indice(2),1);
  y2=nodeCoordinates(indice(2),2);
  z2=nodeCoordinates(indice(2),3);
  L = sqrt((x2-x1)*(x2-x1) + (y2-y1)*(y2-y1) +...
      (z2-z1)*(z2-z1));
  CXx = (x2-x1)/L;CYx = (y2-y1)/L;CZx = (z2-z1)/L;

  T = [CXx*CXx CXx*CYx CXx*CZx ; CYx*CXx CYx*CYx CYx*CZx ; ...
        CZx*CXx CZx*CYx CZx*CZx];
  stiffness(elementDof,elementDof)=...
      stiffness(elementDof,elementDof)+E*A(e)/L*[T -T ; -T T];
end
```

and function stresses3Dtruss.m for computation of stresses at 3D trusses.

```
function stresses3Dtruss(numberElements,elementNodes,...
    nodeCoordinates,displacements,E)

% stresses in 3D truss elements
fprintf('Stresses in elements\n')
ff=zeros(numberElements,6); format
for e=1:numberElements;
  % elementDof: element degrees of freedom (Dof)
  indice=elementNodes(e,:)    ;
  elementDof=[3*indice(1)-2 3*indice(1)-1 3*indice(1)...
          3*indice(2)-2 3*indice(2)-1 3*indice(2)]  ;
  x1=nodeCoordinates(indice(1),1);
  y1=nodeCoordinates(indice(1),2);
  z1=nodeCoordinates(indice(1),3);
  x2=nodeCoordinates(indice(2),1);
  y2=nodeCoordinates(indice(2),2);
  z2=nodeCoordinates(indice(2),3);
  L = sqrt((x2-x1)*(x2-x1) + (y2-y1)*(y2-y1) +...
      (z2-z1)*(z2-z1));
    CXx = (x2-x1)/L;CYx = (y2-y1)/L;CZx = (z2-z1)/L;

  u=displacements(elementDof);
    member_stress(e)=E/L*[-CXx -CYx -CZx CXx CYx CZx]*u;
    fprintf('%3d %12.8f\n',e, member_stress(e));
end
```

Chapter 6
Bernoulli beams

6.1 Introduction

In this chapter we perform the static analysis of Bernoulli beams under bending loads. The beam is defined in the xy plane, with constant cross-section area A (figure 6.1).

The Euler-Bernoulli beam theory assumes that undeformed plane sections remain plane under deformation. The displacement u, at a distance y of the beam middle axis is given by

$$u = -y\frac{dw}{dx} \tag{6.1}$$

where w is the transverse displacement.

Strains are defined as

$$\epsilon_x = \frac{\partial u}{\partial x} = -y\frac{\partial^2 w}{\partial x^2}; \qquad \gamma_{xy} = \frac{\partial u}{\partial y} + \frac{\partial w}{\partial x} = 0 \tag{6.2}$$

The elastic strain deformation is obtained as

$$U = \frac{1}{2}\int_V \sigma_x \epsilon_x dV = \frac{1}{2}\int_V E\epsilon_x^2 dV \tag{6.3}$$

Taking $dV = dAdx$, and integrating in the z direction, we obtain

$$U = \frac{1}{2}\int_{-a}^{a} EI_z \left(\frac{\partial^2 w}{\partial x^2}\right)^2 dx \tag{6.4}$$

where I_z is the second moment of area of the beam cross-section.

The external work for this element is given by

$$\delta W = \int_{-a}^{a} p\delta w dx \tag{6.5}$$

A.J.M. Ferreira, *MATLAB Codes for Finite Element Analysis:*
Solids and Structures, Solid Mechanics and Its Applications 157,
© Springer Science+Business Media B.V. 2009

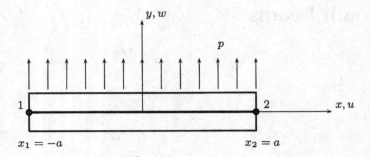

Fig. 6.1 Bernoulli beam element with two nodes

At each node we consider two degrees of freedom, w and $\dfrac{dw}{dx}$,

$$\mathbf{w}^{eT} = \begin{bmatrix} w_1 & \dfrac{dw_1}{dx} & w_2 & \dfrac{dw_2}{dx} \end{bmatrix} \tag{6.6}$$

The transverse displacement is interpolated by Hermite shape functions as

$$\mathbf{w} = \mathbf{N}(\xi)\mathbf{w}^e \tag{6.7}$$

being the shape functions defined as

$$N_1(\xi) = \frac{1}{4}(2 - 3\xi + \xi^2) \tag{6.8}$$

$$N_2(\xi) = \frac{1}{4}(1 - \xi - \xi^2 + \xi^3) \tag{6.9}$$

$$N_3(\xi) = \frac{1}{4}(2 + 3\xi - \xi^2) \tag{6.10}$$

$$N_4(\xi) = \frac{1}{4}(-1 - \xi + \xi^2 + \xi^3) \tag{6.11}$$

The strain energy is obtained as

$$U = \frac{1}{2}\int_{-a}^{a} EI_z \left(\frac{\partial^2 w}{\partial x^2}\right)^2 dx = \frac{1}{2}\int_{-1}^{1} \frac{EI_z}{a^4} \left(\frac{\partial^2 w}{\partial \xi^2}\right)^2 a\,d\xi$$

$$= \frac{1}{2}\mathbf{w}^{eT} \frac{EI_z}{a^3} \int_{-1}^{1} \mathbf{N}''^{T}\mathbf{N}'' d\xi \mathbf{w}^e \tag{6.12}$$

where $\mathbf{N}'' = \dfrac{d^2\mathbf{N}}{d\xi^2}$. The element stiffness matrix is then obtained as

$$\mathbf{K}^e = \frac{EI_z}{a^3} \int_{-1}^{1} \mathbf{N}''^T \mathbf{N}'' d\xi = \frac{EI_z}{2a^3} \begin{bmatrix} 3 & 3a & -3 & 3a \\ 3a & 4a^2 & -3a & 2a^2 \\ -3 & -3a & 3 & -3a \\ 3a & 2a^2 & -3a & 4a^2 \end{bmatrix} \qquad (6.13)$$

The work done by the distributed forces is defined as

$$\delta W^e = \int_{-a}^{a} p\delta w dx = \int_{-1}^{1} p\delta w a d\xi = \delta \mathbf{w}^{eT} a \int_{-1}^{1} p\mathbf{N}^T d\xi \qquad (6.14)$$

The vector of nodal forces equivalent to distributed forces is obtained as

$$\mathbf{f}^e = ap \int_{-1}^{1} \mathbf{N}^T d\xi = \frac{ap}{3} \begin{bmatrix} 3 \\ a \\ 3 \\ -a \end{bmatrix} \qquad (6.15)$$

6.2 Bernoulli beam problem

In figures 6.2 and 6.3 we consider a simply-supported and clamped Bernoulli beam in bending, under uniform load.

Code problem9.m solves this problem for various boundary conditions. The user can define the number of elements. The maximum transverse displacement (w_{max}) for simply-supported beam is 0.0130208 which matches the analytical solution, while for clamped beam $w_{max} = 0.002604$, also matching the analytical solution, as illustrated in figure 6.4.

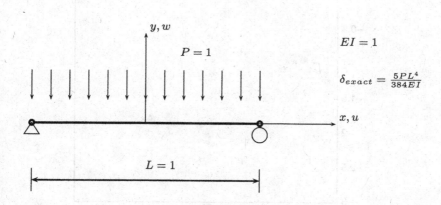

Fig. 6.2 Simply-supported Bernoulli problem, under uniform load, problem9.m

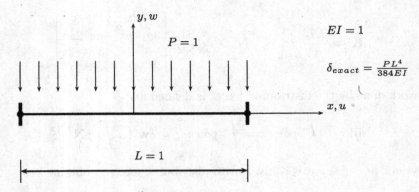

Fig. 6.3 Clamped Bernoulli problem, under uniform load

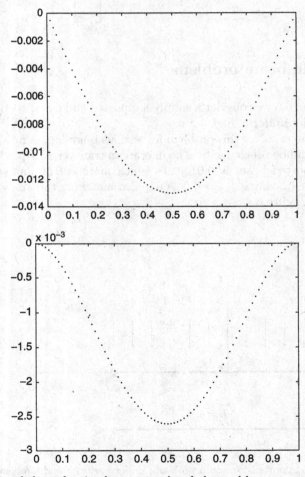

Fig. 6.4 Deformed shape for simply-supported and clamped beams

```
%..........................................................

% MATLAB codes for Finite Element Analysis
% problem9.m
% antonio ferreira 2008

% clear memory
clear all

% E; modulus of elasticity
% I: second moment of area
% L: length of bar
E=1;  I=1;  EI=E*I;

% generation of coordinates and connectivities
numberElements=80;
nodeCoordinates=linspace(0,1,numberElements+1)';
xx=nodeCoordinates;L=max(nodeCoordinates);
numberNodes=size(nodeCoordinates,1);
xx=nodeCoordinates(:,1);
for i=1:numberElements;
    elementNodes(i,1)=i;
    elementNodes(i,2)=i+1;
end

% distributed load
P=-1;

% for structure:
    % displacements: displacement vector
    % force : force vector
    % stiffness: stiffness matrix
    % GDof: global number of degrees of freedom
GDof=2*numberNodes;
U=zeros(GDof,1);

% stiffess matrix and force vector
[stiffness,force]=...
    formStiffnessBernoulliBeam(GDof,numberElements,...
    elementNodes,numberNodes,xx,EI,P);

% boundary conditions and solution
```

```
% clamped-clamped
%fixedNodeU =[1 2*numberElements+1]';
%fixedNodeV =[2 2*numberElements+2]';
% simply supported-simply supported
fixedNodeU =[1 2*numberElements+1]'; fixedNodeV =[]';
% clamped at x=0
%fixedNodeU =[1]'; fixedNodeV =[2]';

prescribedDof=[fixedNodeU;fixedNodeV];
% solution
displacements=solution(GDof,prescribedDof,stiffness,force);

% output displacements/reactions
outputDisplacementsReactions(displacements,stiffness,...
    GDof,prescribedDof)

% drawing deformed shape
U=displacements(1:2:2*numberNodes);
plot(nodeCoordinates,U,'.')
```

This code calls function formStiffnessBernoulliBeam.m for the computation of the stiffness matrix and the force vector for the Bernoulli beam two-node element.

```
%.............................................................

function  [stiffness,force]=...
    formStiffnessBernoulliBeam(GDof,numberElements,...
    elementNodes,numberNodes,xx,EI,P);

force=zeros(GDof,1);
stiffness=zeros(GDof);
% calculation of the system stiffness matrix
% and force vector
for e=1:numberElements;
  % elementDof: element degrees of freedom (Dof)
  indice=elementNodes(e,:)   ;
  elementDof=[ 2*(indice(1)-1)+1 2*(indice(2)-1)...
      2*(indice(2)-1)+1 2*(indice(2)-1)+2];
 % ndof=length(indice);
  % length of element
```

```
    LElem=xx(indice(2))-xx(indice(1))   ;
    ll=LElem;
    k1=EI/(LElem)^3*[12    6*LElem -12 6*LElem;
        6*LElem 4*LElem^2 -6*LElem 2*LElem^2;
        -12 -6*LElem 12 -6*LElem ;
        6*LElem 2*LElem^2 -6*LElem 4*LElem^2];

    f1=[P*LElem/2 P*LElem*LElem/12 P*LElem/2 ...
        -P*LElem*LElem/12]';

    % equivalent force vector
    force(elementDof)=force(elementDof)+f1;

    % stiffness matrix
    stiffness(elementDof,elementDof)=...
        stiffness(elementDof,elementDof)+k1;
end
```

6.3 Bernoulli beam with spring

Figure 6.5 illustrates a beam in bending, clamped at one end and supported by a spring at the other end. Code problem9a.m illustrates the use of MATLAB for solving this problem.

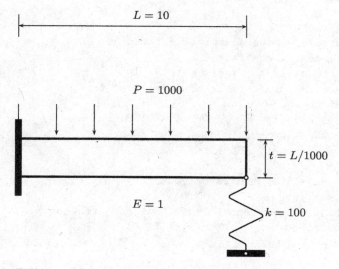

Fig. 6.5 Bernoulli beam with spring, under uniform load, problem9a.m

```
%...............................................................

% MATLAB codes for Finite Element Analysis
% problem9a.m
% antonio ferreira 2008

% clear memory
clear all

% E; modulus of elasticity
% I: second moment of area
% L: length of bar
E=1e6;   L=10;  t=L/1000;I=1*t^3/12;   EI=E*I;

% generation of coordinates and connectivities
numberElements=3;
nodeCoordinates=linspace(0,L,numberElements+1)';
xx=nodeCoordinates;L=max(nodeCoordinates);
for i=1:numberElements;
    elementNodes(i,1)=i;
    elementNodes(i,2)=i+1;
end
numberNodes=size(nodeCoordinates,1);
xx=nodeCoordinates(:,1);

% distributed force
P=-1000;

% for structure:
    % displacements: displacement vector
    % force : force vector
    % stiffness: stiffness matrix
    % GDof: global number of degrees of freedom
GDof=2*numberNodes;
%U=zeros(GDof,1);
stiffnessSpring=zeros(GDof+1);
forceSpring=zeros(GDof+1,1);

% stiffess matrix and force vector
[stiffness,force]=...
    formStiffnessBernoulliBeam(GDof,numberElements,...
    elementNodes,numberNodes,xx,EI,P);
```

```
% spring added
stiffnessSpring(1:GDof,1:GDof)=stiffness;
forceSpring(1:GDof)=force;
k=10;
stiffnessSpring([GDof-1 GDof+1],[GDof-1 GDof+1])=...
    stiffnessSpring([GDof-1 GDof+1],[GDof-1 GDof+1])+[k -k;-k k];

% boundary conditions and solution
fixedNodeU =[1]'; fixedNodeV =[2]';
prescribedDof=[fixedNodeU;fixedNodeV;GDof+1];

% solution
displacements=solution(GDof+1,prescribedDof,...
    stiffnessSpring,forceSpring);

% displacements
disp('Displacements')
jj=1:GDof+1; format
[jj' displacements]

% drawing deformed shape
U=displacements(1:2:2*numberNodes);
plot(nodeCoordinates,U,'*')

% exact solution by Bathe (Solutions Manual of Procedures ...)
load=[L*P/3;L*P/3;L*P/6];
K=E*I/L^3*[189 -108 27;-108 135 -54;27 -54 27+k*L^3/E/I];
X=K\load
```

Results are compared with finite element solution by Bathe [12] in his Solution
Manual. The transverse displacement at the right end of the beam is 0.0037e5 for
the MATLAB solution, while Bathe presents 0.0039e5, using three finite elements.

Chapter 7
2D frames

7.1 Introduction

In this chapter bidimensional frames under static loading are analysed. In figure 7.1, we show the 2D frame two-noded element. Each node has three global degrees of freedom, two displacements in global axes and one rotation.

The vector of global displacements is given by

$$\mathbf{u}^T = [u_1, \quad u_2, \quad u_3, \quad u_4, \quad u_5, \quad u_6] \tag{7.1}$$

We define a local basis with cosines l, m, with respect to θ, the angle between x' and x. In this local coordinate set the displacements are detailed as

$$\mathbf{u'}^T = [u'_1, \quad u'_2, \quad u'_3, \quad u'_4, \quad u'_5, \quad u'_6] \tag{7.2}$$

Noting that $u'_3 = u_3, u'_6 = u_6$, we derive a local-global transformation matrix in the form

$$\mathbf{u'} = \mathbf{L}\mathbf{u} \tag{7.3}$$

where

$$\mathbf{L} = \begin{bmatrix} l & m & 0 & 0 & 0 & 0 \\ -m & l & 0 & 0 & 0 & 0 \\ 0 & 0 & 1 & 0 & 0 & 0 \\ 0 & 0 & 0 & l & m & 0 \\ 0 & 0 & 0 & -m & l & 0 \\ 0 & 0 & 0 & 0 & 0 & 1 \end{bmatrix} \tag{7.4}$$

In local coordinates, the stiffness matrix of the frame element is obtained by combination of the stiffness of the bar element and the Bernoulli beam element, in the form

A.J.M. Ferreira, *MATLAB Codes for Finite Element Analysis:*
Solids and Structures, Solid Mechanics and Its Applications 157,
© Springer Science+Business Media B.V. 2009

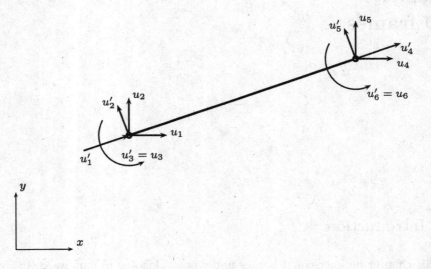

Fig. 7.1 A 2D frame element

$$\mathbf{K}'^e = \begin{bmatrix} \dfrac{EA}{L} & 0 & 0 & -\dfrac{EA}{L} & 0 & 0 \\[2.2ex] & \dfrac{12EI}{L^3} & \dfrac{6EI}{L^2} & 0 & \dfrac{12EI}{L^3} & \dfrac{6EI}{L^2} \\[2.2ex] & & \dfrac{4EI}{L} & 0 & -\dfrac{6EI}{L^2} & \dfrac{2EI}{L} \\[2.2ex] & & & \dfrac{EA}{L} & 0 & 0 \\[2.2ex] & & & & \dfrac{12EI}{L^3} & -\dfrac{6EI}{L^2} \\[2.2ex] sim. & & & & & \dfrac{4EI}{L} \end{bmatrix} \tag{7.5}$$

In global coordinates, the strain energy is given by

$$U_e = \frac{1}{2}\mathbf{u}'^T\mathbf{K}'\mathbf{u}' = \frac{1}{2}\mathbf{u}^T\mathbf{L}^T\mathbf{K}'\mathbf{L}\mathbf{u} = \mathbf{u}^T\mathbf{K}\mathbf{u} \tag{7.6}$$

where

$$\mathbf{K} = \mathbf{L}^T\mathbf{K}'\mathbf{L} \tag{7.7}$$

7.2 An example of 2D frame

Consider the bidimensional frame illustrated in figure 7.2. The code for solving this problem is problem10.m. The degrees of freedom are shown in figure 7.3. Note that the new numbering of global degrees of freedom, to exploit MATLAB programming strengths. In order to match the ordering of degrees of freedom, the stiffness matrix has to be rearranged, as shown in the code listing.

```
%.......................................................................
% MATLAB codes for Finite Element Analysis
% problem10.m
```

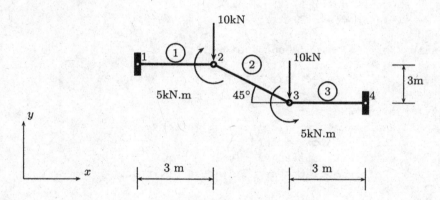

Fig. 7.2 A 2D frame example, problem10.m

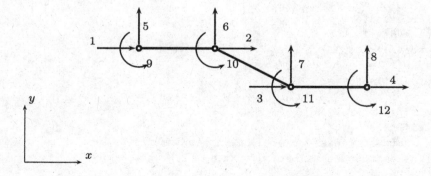

Fig. 7.3 Degrees of freedom for problem 10

```
% antonio ferreira 2008

% clear memory
clear all

% E; modulus of elasticity
% I: second moment of area
% L: length of bar
E=210000; A=100; I=2e8; EA=E*A; EI=E*I;

% generation of coordinates and connectivities
numberElements=3;
p1=3000*(1+cos(pi/4));
nodeCoordinates=[0 3000;3000 3000;p1 0;p1+3000 0];
xx=nodeCoordinates;
for i=1:numberElements;
    elementNodes(i,1)=i;
    elementNodes(i,2)=i+1;
end
numberNodes=size(nodeCoordinates,1);
xx=nodeCoordinates(:,1);
yy=nodeCoordinates(:,2);

% for structure:
    % displacements: displacement vector
    % force : force vector
    % stiffness: stiffness matrix
    % GDof: global number of degrees of freedom
GDof=3*numberNodes;
U=zeros(GDof,1);
force=zeros(GDof,1);

%force vector
force(6)=-10000;
force(7)=-10000;
force(10)=-5e6;
force(11)=5e6;

% stiffness matrix
[stiffness]=...
    formStiffness2Dframe(GDof,numberElements,...
    elementNodes,numberNodes,xx,yy,EI,EA);
```

```
% boundary conditions and solution
prescribedDof=[1 4 5 8 9 12]';

% solution
displacements=solution(GDof,prescribedDof,stiffness,force);

% output displacements/reactions
outputDisplacementsReactions(displacements,stiffness,...
    GDof,prescribedDof)

% drawing undeformed and deformed meshes
U=displacements;
clf
drawingMesh(nodeCoordinates+500*[U(1:numberNodes)...
    U(numberNodes+1:2*numberNodes)],elementNodes,'L2','k.-');
drawingMesh(nodeCoordinates,elementNodes,'L2','k--');
```

Code problem10.m calls function formStiffness2Dframe.m, to compute the stiffness matrix of bidimensional frame elements.

```
function  [stiffness]=...
    formStiffness2Dframe(GDof,numberElements,...
    elementNodes,numberNodes,xx,yy,EI,EA);

stiffness=zeros(GDof);
% computation of the system stiffness matrix
for e=1:numberElements;
  % elementDof: element degrees of freedom (Dof)
  indice=elementNodes(e,:)    ;
  elementDof=[ indice indice+numberNodes indice+2*numberNodes] ;
  nn=length(indice);
  xa=xx(indice(2))-xx(indice(1))
  ya=yy(indice(2))-yy(indice(1));
  length_element=sqrt(xa*xa+ya*ya);
  cosa=xa/length_element;
  sena=ya/length_element;
  ll=length_element;
```

```
 L= [cosa*eye(2) sena*eye(2) zeros(2);
     -sena*eye(2) cosa*eye(2) zeros(2);
     zeros(2,4) eye(2)];

oneu=[1 -1;-1 1];
oneu2=[1 -1;1 -1];
oneu3=[1 1;-1 -1];
oneu4=[4 2;2 4];

k1=[EA/ll*oneu    zeros(2,4);
    zeros(2) 12*EI/ll^3*oneu 6*EI/ll^2*oneu3;
    zeros(2) 6*EI/ll^2*oneu2 EI/ll*oneu4];

    stiffness(elementDof,elementDof)=...
        stiffness(elementDof,elementDof)+L'*k1*L;
end
```

Results are given as:

Displacements

ans =

```
    1.0000          0
    2.0000     0.0000
    3.0000    -0.0000
    4.0000          0
    5.0000          0
    6.0000    -1.3496
    7.0000    -1.3496
    8.0000          0
    9.0000          0
   10.0000    -0.0005
   11.0000     0.0005
   12.0000          0
```

reactions

ans =

```
   1.0e+07 *

     0.0000    -0.0000
```

Fig. 7.4 Deformed shape
of problem 10

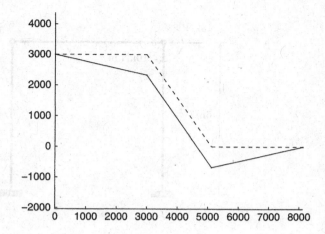

```
0.0000      0.0000
0.0000      0.0010
0.0000      0.0010
0.0000      2.2596
0.0000     -2.2596
```

The deformed shape is given in figure 7.4.

7.3 Another example of 2D frame

Consider the frame illustrated in figure 7.5. The code for solving this problem is
problem11.m. The degrees of freedom are shown in figure 7.6.

```
%.............................................................

% MATLAB codes for Finite Element Analysis
% problem11.m
% 2D frame
% antonio ferreira 2008

% clear memory
clear all

% E; modulus of elasticity
% I: second moment of area
```

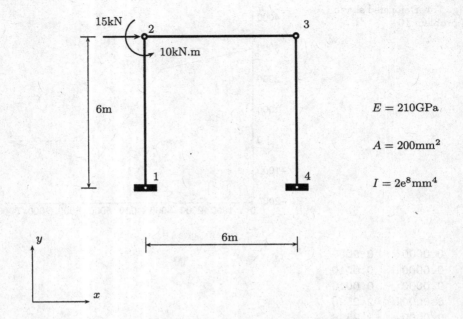

Fig. 7.5 A 2D frame example: geometry, materials and loads, problem11.m

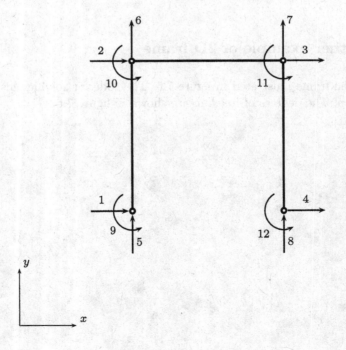

Fig. 7.6 A 2D frame example: degree of freedom ordering

```
% L: length of bar
E=210000; A=200; I=2e8; EA=E*A; EI=E*I;

% generation of coordinates and connectivities
numberElements=3;
nodeCoordinates=[0 0;0 6000;6000 6000;6000 0];
xx=nodeCoordinates;
for i=1:numberElements;
    elementNodes(i,1)=i;
    elementNodes(i,2)=i+1;
end
numberNodes=size(nodeCoordinates,1);
xx=nodeCoordinates(:,1);
yy=nodeCoordinates(:,2);

% for structure:
    % displacements: displacement vector
    % force : force vector
    % stiffness: stiffness matrix
    % GDof: global number of degrees of freedom
GDof=3*numberNodes;
U=zeros(GDof,1);
force=zeros(GDof,1);

%force vector
force(2)=15000;
force(10)=10e6;

% stiffness matrix
[stiffness]=...
    formStiffness2Dframe(GDof,numberElements,...
    elementNodes,numberNodes,xx,yy,EI,EA);

% boundary conditions and solution
prescribedDof=[1 4 5 8 9 12]';

% solution
displacements=solution(GDof,prescribedDof,stiffness,force);

% output displacements/reactions
outputDisplacementsReactions(displacements,stiffness,...
    GDof,prescribedDof)
```

```
%drawing mesh and deformed shape
clf
U=displacements;
drawingMesh(nodeCoordinates+500*[U(1:numberNodes)...
    U(numberNodes+1:2*numberNodes)],elementNodes,'L2','k.-');
drawingMesh(nodeCoordinates,elementNodes,'L2','k--');
```

Results are listed below.

Displacements

ans =

```
    1.0000          0
    2.0000     5.2843
    3.0000     4.4052
    4.0000          0
    5.0000          0
    6.0000     0.6522
    7.0000    -0.6522
    8.0000          0
    9.0000          0
   10.0000    -0.0005
   11.0000    -0.0006
   12.0000          0
```

reactions

ans =

   1.0e+07 *

```
    0.0000    -0.0009
    0.0000    -0.0006
    0.0000    -0.0005
    0.0000     0.0005
    0.0000     3.0022
    0.0000     2.2586
```

The deformed shape of this problem is illustrated in figure 7.7.

Increasing the number of elements will give us a more realistic deformed shape. Code problem11b.m considers 12 elements.

Fig. 7.7 Deformed shape
of problem 11

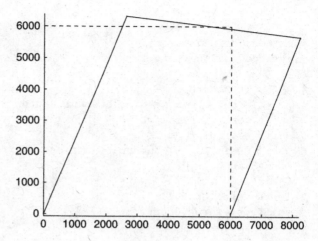

```
%.............................................................

% MATLAB codes for Finite Element Analysis
% problem11b.m
% 2D frame
% antonio ferreira 2008

% clear memory
clear all

% E; modulus of elasticity
% I: second moment of area
% L: length of bar
E=210000; A=200; I=2e8; EA=E*A; EI=E*I;

% generation of coordinates and connectivities
numberElements=12;
nodeCoordinates=[0 0;0 1500;0 3000;0 4500 ;
                 0 6000;1500 6000;3000 6000;
                 4500 6000;6000 6000;6000 4500;
                 6000 3000;6000 1500;6000 0];
xx=nodeCoordinates;
for i=1:numberElements;
    elementNodes(i,1)=i;
    elementNodes(i,2)=i+1;
end
numberNodes=size(nodeCoordinates,1);
```

```
xx=nodeCoordinates(:,1);
yy=nodeCoordinates(:,2);


% for structure:
    % displacements: displacement vector
    % force : force vector
    % stiffness: stiffness matrix
    % GDof: global number of degrees of freedom
GDof=3*numberNodes;
U=zeros(GDof,1);
force=zeros(GDof,1);
stiffness=zeros(GDof);


%force vector
force(5)=15000;
force(31)=10e6;


% stiffness matrix
[stiffness]=...
    formStiffness2Dframe(GDof,numberElements,...
    elementNodes,numberNodes,xx,yy,EI,EA);


% boundary conditions and solution
prescribedDof=[1 13 14 26 27 39]';


% solution
displacements=solution(GDof,prescribedDof,stiffness,force);


% output displacements/reactions
outputDisplacementsReactions(displacements,stiffness,...
    GDof,prescribedDof)


%drawing mesh and deformed shape
U=displacements;
clf
drawingMesh(nodeCoordinates+500*[U(1:numberNodes)...
    U(numberNodes+1:2*numberNodes)],elementNodes,'L2','k.-');
drawingMesh(nodeCoordinates,elementNodes,'L2','k--');
```

Important to note the generation of node coordinates and element nodes, also the force numbering as well as the boundary conditions ordering. As before we consider first the u degrees of freedom, then the v degrees of freedom and finally rotations.
Results are given below.

Displacements

ans =

1.0000	0
2.0000	0.6857
3.0000	2.2689
4.0000	4.0387
5.0000	5.2843
6.0000	5.0645
7.0000	4.8447
8.0000	4.6249
9.0000	4.4052
10.0000	3.2197
11.0000	1.7606
12.0000	0.5226
13.0000	0
14.0000	0
15.0000	0.1630
16.0000	0.3261
17.0000	0.4891
18.0000	0.6522
19.0000	0.1942
20.0000	0.0687
21.0000	−0.0912
22.0000	−0.6522
23.0000	−0.4891
24.0000	−0.3261
25.0000	−0.1630
26.0000	0
27.0000	0
28.0000	−0.0008
29.0000	−0.0012
30.0000	−0.0011
31.0000	−0.0005
32.0000	−0.0002
33.0000	−0.0001
34.0000	−0.0002
35.0000	−0.0006
36.0000	−0.0009
37.0000	−0.0010

```
  38.0000   -0.0006
  39.0000        0
```

reactions

ans =

```
  1.0e+07 *

  0.0000   -0.0009
  0.0000   -0.0006
  0.0000   -0.0005
  0.0000    0.0005
  0.0000    3.0022
  0.0000    2.2586
```

>>

The deformed shape of this problem is illustrated in figure 7.8.

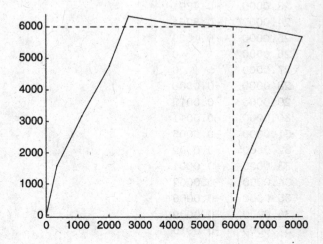

Fig. 7.8 Deformed shape
of problem 11b

Chapter 8
Analysis of 3D frames

8.1 Introduction

The analysis of three dimensional frames is quite similar to the analysis of 2D frames. In the two-node 3D frame finite element we now consider in each node three displacements and three rotations with respect to the three global cartesian axes.

8.2 Stiffness matrix and vector of equivalent nodal forces

In the local coordinate system, the stiffness matrix is given by

$$
K'^e =
\begin{bmatrix}
\frac{EA}{L} & 0 & 0 & 0 & 0 & 0 & -\frac{EA}{L} & 0 & 0 & 0 & 0 & 0 \\
 & \frac{12EI_z}{L^3} & 0 & 0 & 0 & \frac{6EI_z}{L^2} & 0 & -\frac{12EI_z}{L^3} & 0 & 0 & 0 & \frac{6EI_z}{L^2} \\
 & & \frac{12EI_y}{L^3} & 0 & -\frac{6EI_y}{L^2} & 0 & 0 & 0 & -\frac{12EI_y}{L^3} & 0 & -\frac{6EI_y}{L^2} & 0 \\
 & & & \frac{GJ}{L} & 0 & 0 & 0 & 0 & 0 & -\frac{GJ}{L} & 0 & 0 \\
 & & & & \frac{4EI_y}{L} & 0 & 0 & 0 & \frac{6EI_y}{L^2} & 0 & \frac{2EI_y}{L} & 0 \\
 & & & & & \frac{4EI_z}{L} & 0 & -\frac{6EI_z}{L^2} & 0 & 0 & 0 & \frac{2EI_z}{L} \\
 & & & & & & \frac{EA}{L} & 0 & 0 & 0 & 0 & 0 \\
 & & & & & & & \frac{12EI_z}{L^3} & 0 & 0 & 0 & \frac{-6EI_z}{L^2} \\
 & & & & & & & & \frac{12EI_y}{L^3} & 0 & \frac{6EI_y}{L^2} & 0 \\
 & & & & & & & & & \frac{GJ}{L} & 0 & 0 \\
 & & & & & & & & & & \frac{4EI_y}{L} & 0 \\
 sim. & & & & & & & & & & & \frac{4EI_z}{L}
\end{bmatrix}
\tag{8.1}
$$

A.J.M. Ferreira, *MATLAB Codes for Finite Element Analysis:*
Solids and Structures, Solid Mechanics and Its Applications 157,
© Springer Science+Business Media B.V. 2009

After transformation to the global axes, the stiffness matrix in global coordinates is obtained as

$$\mathbf{K} = \mathbf{R}^T\mathbf{K}'\mathbf{R}$$

where the rotation matrix R is defined as

$$\mathbf{R} = \begin{bmatrix} r & 0 & 0 & 0 \\ 0 & r & 0 & 0 \\ 0 & 0 & r & 0 \\ 0 & 0 & 0 & r \end{bmatrix} \tag{8.2}$$

being

$$\mathbf{r} = \begin{bmatrix} C_{Xx} & C_{Yx} & C_{Zx} \\ C_{Xy} & C_{Yy} & C_{Zy} \\ C_{Xz} & C_{Yz} & C_{Zz} \end{bmatrix} \tag{8.3}$$

and

$$C_{Xx} = cos\theta_{Xx}$$

where angles θ_{Xx}, θ_{Yx}, and θ_{Zx}, are measured from global axes X, Y, and Z, with respect to the local axis x, respectively. The two-node 3D frame element has six degrees of freedom per node. Given the nodal displacements, it is possible to calculate the reactions at the supports by

$$\mathbf{F} = \mathbf{KU} \tag{8.4}$$

where $\mathbf{K}$ and $\mathbf{U}$ are the structure stiffness matrix and the vector of nodal displacement, respectively. The element forces are also evaluated by axes transformation as

$$\mathbf{f}_e = \mathbf{k}_e\mathbf{R}\mathbf{U}^e \tag{8.5}$$

8.3 First 3D frame example

The first 3D frame example is illustrated in figure 8.1. We consider $E = 210$ GPa, $G = 84$ GPa, $A = 2 \times 10^{-2}$ m^2, $I_y = 10 \times 10^{-5}$ m^4, $I_z = 20 \times 10^{-5}$ m^4, $J = 5 \times 10^{-5}$ m^4.

Code problem12.m solves this problem, and calls function formStiffness3D frame.m, that computes the stiffness matrix of the 3D frame element.

```
%............................................................

% MATLAB codes for Finite Element Analysis
% problem12.m
% antonio ferreira 2008
```

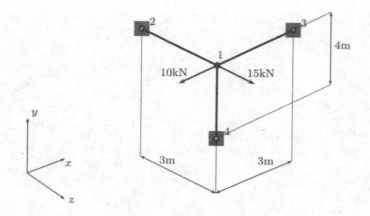

Fig. 8.1 A 3D frame example (problem12.m)

```
% clear memory
clear all

% E; modulus of elasticity
% I: second moment of area
% L: length of bar
E=210e6; A=0.02;
Iy=10e-5;   Iz=20e-5; J=5e-5; G=84e6;

% generation of coordinates and connectivities
nodeCoordinates=[0 0 0;    3 0 0; 0 0 -3; 0 -4 0];
xx=nodeCoordinates(:,1);
yy=nodeCoordinates(:,2);
zz=nodeCoordinates(:,3);
elementNodes=[1 2;1 3;1 4];
numberNodes=size(nodeCoordinates,1);
numberElements=size(elementNodes,1);

% for structure:
    % displacements: displacement vector
    % force : force vector
    % stiffness: stiffness matrix
    % GDof: global number of degrees of freedom
GDof=6*numberNodes;
U=zeros(GDof,1);
force=zeros(GDof,1);
stiffness=zeros(GDof,GDof);
```

```
%force vector
force(1)=-10;
force(3)=20;

% stiffness matrix
[stiffness]=...
    formStiffness3Dframe(GDof,numberElements,...
    elementNodes,numberNodes,nodeCoordinates,E,A,Iz,Iy,G,J);

% boundary conditions and solution
prescribedDof=[7:24];

% solution
displacements=solution(GDof,prescribedDof,stiffness,force)

% displacements
disp('Displacements')
jj=1:GDof; format long
f=[jj; displacements'];
fprintf('node U\n')
fprintf('%3d %12.8f\n',f)
```

Listing of formStiffness3Dframe.m:

```
function  [stiffness]=...
    formStiffness3Dframe(GDof,numberElements,...
    elementNodes,numberNodes,nodeCoordinates,E,A,Iz,Iy,G,J);

stiffness=zeros(GDof);
% computation of the system stiffness matrix
for e=1:numberElements;
  % elementDof: element degrees of freedom (Dof)
  indice=elementNodes(e,:)    ;
  elementDof=[6*indice(1)-5 6*indice(1)-4 6*indice(1)-3 ...
              6*indice(1)-2 6*indice(1)-1 6*indice(1)...
              6*indice(2)-5 6*indice(2)-4 6*indice(2)-3 ...
              6*indice(2)-2 6*indice(2)-1 6*indice(2)] ;
  x1=nodeCoordinates(indice(1),1);
  y1=nodeCoordinates(indice(1),2);
  z1=nodeCoordinates(indice(1),3);
  x2=nodeCoordinates(indice(2),1);
```

```
   y2=nodeCoordinates(indice(2),2);
   z2=nodeCoordinates(indice(2),3);

L = sqrt((x2-x1)*(x2-x1) + (y2-y1)*(y2-y1) +...
         (z2-z1)*(z2-z1));
k1 = E*A/L;
k2 = 12*E*Iz/(L*L*L);
k3 = 6*E*Iz/(L*L);
k4 = 4*E*Iz/L;
k5 = 2*E*Iz/L;
k6 = 12*E*Iy/(L*L*L);
k7 = 6*E*Iy/(L*L);
k8 = 4*E*Iy/L;
k9 = 2*E*Iy/L;
k10 = G*J/L;

a=[k1 0 0; 0 k2 0; 0 0 k6];
b=[ 0 0 0;0 0 k3; 0 -k7 0];
c=[k10 0 0;0 k8 0; 0 0 k4];
d=[-k10 0 0;0 k9 0;0 0 k5];

k = [a b -a b;b' c b d; (-a)' b' a -b;b' d' (-b)' c];

if x1 == x2 & y1 == y2
   if z2 > z1
      Lambda = [0 0 1 ; 0 1 0 ; -1 0 0];
   else
      Lambda = [0 0 -1 ; 0 1 0 ; 1 0 0];
   end
else
   CXx = (x2-x1)/L;
CYx = (y2-y1)/L;
CZx = (z2-z1)/L;
D = sqrt(CXx*CXx + CYx*CYx);
CXy = -CYx/D;
CYy = CXx/D;
CZy = 0;
CXz = -CXx*CZx/D;
CYz = -CYx*CZx/D;
CZz = D;
Lambda = [CXx CYx CZx ;CXy CYy CZy ;CXz CYz CZz];

end
```

```
R = [Lambda zeros(3,9); zeros(3) Lambda zeros(3,6);
    zeros(3,6) Lambda zeros(3);zeros(3,9) Lambda];

    stiffness(elementDof,elementDof)=...
        stiffness(elementDof,elementDof)+R'*k*R;
end
```

Results are listed as follows.

```
Displacements
node U
   1  -0.00000705
   2  -0.00000007
   3   0.00001418
   4   0.00000145
   5   0.00000175
   6   0.00000114
   7   0.00000000
   8   0.00000000
   9   0.00000000
  10   0.00000000
  11   0.00000000
  12   0.00000000
  13   0.00000000
  14   0.00000000
  15   0.00000000
  16   0.00000000
  17   0.00000000
  18   0.00000000
  19   0.00000000
  20   0.00000000
  21   0.00000000
  22   0.00000000
  23   0.00000000
  24   0.00000000
```

8.4 Second 3D frame example

The next 3D problem is illustrated in figure 8.2 and considers $E = 210$ GPa, $G = 84$ GPa, $A = 2 \times 10^{-2}$ m^2, $I_y = 10 \times 10^{-5}$ m^4, $I_z = 20 \times 10^{-5}$ m^4, $J = 5 \times 10^{-5}$ m^4. The MATLAB code for this problem is problem13.m.

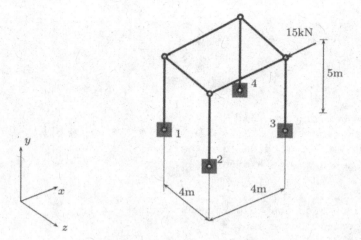

Fig. 8.2 A second 3D frame example (problem13.m)

```
%.........................................................

% MATLAB codes for Finite Element Analysis
% problem13.m
% antonio ferreira 2008

% clear memory
clear all

% E; modulus of elasticity
% I: second moments of area
% J: polar moment of inertia
% G: shear modulus
% L: length of bar
E=210e6; A=0.02;
Iy=10e-5;   Iz=20e-5; J=5e-5; G=84e6;

% generation of coordinates and connectivities
nodeCoordinates=[0 0 0;
    0 0 4;
    4 0 4;
    4 0 0;
    0 5 0;
    0 5 4;
```

```
    4 5 4;
    4 5 0;
    ];
xx=nodeCoordinates(:,1);
yy=nodeCoordinates(:,2);
zz=nodeCoordinates(:,3);
elementNodes=[1 5;2 6;3 7; 4 8; 5 6; 6 7; 7 8; 8 5];
numberNodes=size(nodeCoordinates,1);
numberElements=size(elementNodes,1);

% for structure:
    % displacements: displacement vector
    % force : force vector
    % stiffness: stiffness matrix
    % GDof: global number of degrees of freedom
GDof=6*numberNodes;
U=zeros(GDof,1);
force=zeros(GDof,1);
stiffness=zeros(GDof);

%force vector
force(37)=-15;

% calculation of the system stiffness matrix
% and force vector
% stiffness matrix
[stiffness]=...
    formStiffness3Dframe(GDof,numberElements,...
    elementNodes,numberNodes,nodeCoordinates,E,A,Iz,Iy,G,J);

% boundary conditions and solution
prescribedDof=[1:24];

% solution
displacements=solution(GDof,prescribedDof,stiffness,force);

% displacements
disp('Displacements')
jj=1:GDof; format long
f=[jj; displacements'];
fprintf('node U\n')
fprintf('%3d %12.8f\n',f)
```

```
%drawing mesh and deformed shape
U=displacements;
clf
drawingMesh(nodeCoordinates+500*[U(1:6:6*numberNodes)...
    U(2:6:6*numberNodes) U(3:6:6*numberNodes)],...
    elementNodes,'L2','k.-');
drawingMesh(nodeCoordinates,elementNodes,'L2','k--');
```

Results are obtained as:

Displacements
node U
1	0.00000000
2	0.00000000
3	0.00000000
4	0.00000000
5	0.00000000
6	0.00000000
7	0.00000000
8	0.00000000
9	0.00000000
10	0.00000000
11	0.00000000
12	0.00000000
13	0.00000000
14	0.00000000
15	0.00000000
16	0.00000000
17	0.00000000
18	0.00000000
19	0.00000000
20	0.00000000
21	0.00000000
22	0.00000000
23	0.00000000
24	0.00000000
25	-0.00227293
26	-0.00000393
27	0.00115845
28	0.00007082
29	0.00013772
30	0.00023748
31	-0.00106188
32	-0.00000616

Fig. 8.3 Deformed shape
for problem 13

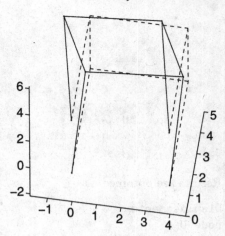

33	0.00115835
34	0.00007201
35	0.00064608
36	0.00011324
37	-0.00106724
38	0.00000581
39	-0.00116612
40	-0.00007274
41	0.00075248
42	0.00011504
43	-0.00227804
44	0.00000679
45	-0.00116139
46	-0.00007078
47	0.00064378
48	0.00023789

The deformed shape of this structure is illustrated in figure 8.3.

Chapter 9
Analysis of grids

9.1 Introduction

In this chapter we perform the static analysis of grids, which are planar structures where forces are applied normal to the grid plane.

At each node a transverse displacement and two rotations are assigned. The stiffness matrix in local cartesian axes is given by

$$\mathbf{k}_e = \begin{bmatrix} \dfrac{12EI}{l_e^3} & 0 & \dfrac{6EI}{l_e^2} & -\dfrac{12EI}{l_e^3} & 0 & \dfrac{6EI}{l_e^2} \\ 0 & \dfrac{GJ}{l_e} & 0 & 0 & -\dfrac{GJ}{l_e} & 0 \\ \dfrac{6EI}{l_e^2} & 0 & \dfrac{4EI}{l_e} & -\dfrac{6EI}{l_e^2} & 0 & \dfrac{2EI}{l_e} \\ -\dfrac{12EI}{l_e^3} & 0 & -\dfrac{6EI}{l_e^2} & \dfrac{12EI}{l_e^3} & 0 & \dfrac{6EI}{l_e^2} \\ 0 & -\dfrac{GJ}{l_e} & 0 & 0 & \dfrac{GJ}{l_e} & 0 \\ \dfrac{6EI}{l_e^2} & 0 & \dfrac{2EI}{l_e} & -\dfrac{6EI}{l_e^2} & 0 & \dfrac{4EI}{l_e} \end{bmatrix} \tag{9.1}$$

where E is the modulus of elasticity, I is the second moment of area, J the polar moment of inertia, and G the shear modulus. The element length is denoted by $L = l_e$.

We consider direction cosines $C = \cos\theta$ and $S = \sin\theta$, being θ the angle between global axis X and local axis x. The rotation matrix is defined as

$$\mathbf{R} = \begin{bmatrix} 1 & 0 & 0 & 0 & 0 & 0 \\ 0 & C & S & 0 & 0 & 0 \\ 0 & -S & C & 0 & 0 & 0 \\ 0 & 0 & 0 & 1 & 0 & 0 \\ 0 & 0 & 0 & 0 & C & S \\ 0 & 0 & 0 & 0 & -S & C \end{bmatrix} \tag{9.2}$$

A.J.M. Ferreira, *MATLAB Codes for Finite Element Analysis:*
Solids and Structures, Solid Mechanics and Its Applications 157,
© Springer Science+Business Media B.V. 2009

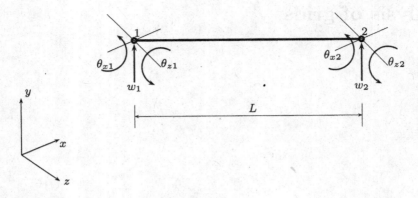

Fig. 9.1 A typical two-node grid element

The stiffness matrix in global cartesian axes is obtained as

$$\mathbf{K}_e = \mathbf{R}^T \mathbf{k}_e \mathbf{R} \tag{9.3}$$

Six degrees of freedom are linked to every grid element, as illustrated in figure 9.1. After computing displacements in global coordinate set, we.compute reactions by

$$\mathbf{F} = \mathbf{K}\mathbf{U} \tag{9.4}$$

where $\mathbf{K}$ and $\mathbf{U}$ is the stiffness matrix and the vector of nodal displacements of the structure, respectively. Element forces are also possible to compute by transformation

$$\mathbf{f}_e = \mathbf{k}_e \mathbf{R} \mathbf{U}^e \tag{9.5}$$

The code for computation of the stiffness matrix of the grid element is listed below.

```
%............................................................
function stiffness=formStiffnessGrid(GDof,...
    numberElements,elementNodes,xx,yy,E,I,G,J)

% function to form global stiffness for grid element
stiffness=zeros(GDof);
for e=1:numberElements;
  % elementDof: element degrees of freedom (Dof)
  indice=elementNodes(e,:)   ;
  elementDof=[...
      (indice(1)-1)*3+1 (indice(1)-1)*3+2 (indice(1)-1)*3+3 ...
      (indice(2)-1)*3+1 (indice(2)-1)*3+2 (indice(2)-1)*3+3] ;
```

```
   xa=xx(indice(2))-xx(indice(1));
   ya=yy(indice(2))-yy(indice(1));
   L=sqrt(xa*xa+ya*ya);
   C=xa/L;
   S=ya/L;

a1 = 12*E*I/(L*L*L);
a2 = 6*E*I/(L*L);
a3 = G*J/L;
a4 = 4*E*I/L;
a5 = 2*E*I/L;
% stiffness in local axes
k = [a1 0 a2 -a1 0 a2 ; 0 a3 0 0 -a3 0 ;
     a2 0 a4 -a2 0 a5 ; -a1 0 -a2 a1 0 -a2 ;
     0 -a3 0 0 a3 0; a2 0 a5 -a2 0 a4];

% transformation matrix
a=[1 0 0; 0 C S;0 -S C];
R=[a zeros(3);zeros(3) a];

   stiffness(elementDof,elementDof)=...
       stiffness(elementDof,elementDof)+R'*k*R;
end
```

The code for computing the forces in elements is listed below.

```
%.......................................................................

function EF=forcesInElementGrid(numberElements,...
    elementNodes,xx,yy,E,I,G,J,displacements)

% forces in elements
EF=zeros(6,numberElements);

for e=1:numberElements;
  % elementDof: element degrees of freedom (Dof)
  indice=elementNodes(e,:)   ;
  elementDof=...
      [(indice(1)-1)*3+1 (indice(1)-1)*3+2 (indice(1)-1)*3+3 ...
      (indice(2)-1)*3+1 (indice(2)-1)*3+2 (indice(2)-1)*3+3] ;
  xa=xx(indice(2))-xx(indice(1));
```

```
   ya=yy(indice(2))-yy(indice(1));
   L=sqrt(xa*xa+ya*ya);
   C=xa/L;
   S=ya/L;

a1 = 12*E*I/(L*L*L);
a2 = 6*E*I/(L*L);
a3 = G*J/L;
a4 = 4*E*I/L;
a5 = 2*E*I/L;

% stiffness in local axes
k = [a1 0 a2 -a1 0 a2 ; 0 a3 0 0 -a3 0 ;
     a2 0 a4 -a2 0 a5 ; -a1 0 -a2 a1 0 -a2 ;
     0 -a3 0 0 a3 0; a2 0 a5 -a2 0 a4];

% transformation matrix
a=[1 0 0; 0 C S;0 -S C];
R=[a zeros(3);zeros(3) a];

% forces in element
EF (:,e)= k*R* displacements(elementDof);
end
```

9.2 A first grid example

The first grid problem is illustrated in figure 9.2. The grid is built from two elements, as illustrated. Given $E = 210$ GPa, $G = 84$ GPa, $I = 20 \times 10^{-5}$ m^4, $J = 5 \times 10^{-5}$ m^4, the MATLAB problem14.m computes displacements, reactions and stresses.

```
%.......................................................................

% MATLAB codes for Finite Element Analysis
% problem14.m
% antonio ferreira 2008

% clear memory
clear all
```

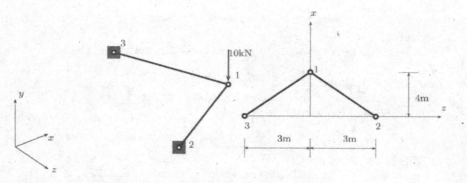

Fig. 9.2 A first grid example, problem14.m

```
% E; modulus of elasticity
% I: second moments of area
% J: polar moment of inertia
% G: shear modulus
% L: length of bar
E=210e6; G=84e6;   I=20e-5;   J=5e-5;

% generation of coordinates and connectivities
nodeCoordinates=[4 0; 0 3; 0 -3];
xx=nodeCoordinates(:,1);
yy=nodeCoordinates(:,2);
elementNodes=[1 2; 3 1];
numberNodes=size(nodeCoordinates,1);
numberElements=size(elementNodes,1);

% GDof: global number of degrees of freedom
GDof=3*numberNodes;
force=zeros(GDof,1);

%force vector
force(1)=-10;

% calculation of the system stiffness matrix
stiffness=formStiffnessGrid(GDof,numberElements,...
    elementNodes,xx,yy,E,I,G,J);

% boundary conditions
prescribedDof=[4:9]';
```

```
% solution
displacements=solution(GDof,prescribedDof,stiffness,force);

% output displacements/reactions
outputDisplacementsReactions(displacements,stiffness,...
    GDof,prescribedDof)

% % forces in elements
disp('forces in elements ')
EF=forcesInElementGrid(numberElements,elementNodes,...
    xx,yy,E,I,G,J,displacements)
```

Results for displacements, reactions and forces in elements are listed below.

```
Displacements

ans =

    1.0000    -0.0048
    2.0000          0
    3.0000    -0.0018
    4.0000          0
    5.0000          0
    6.0000          0
    7.0000          0
    8.0000          0
    9.0000          0

reactions

ans =

    4.0000     5.0000
    5.0000    13.8905
    6.0000    20.0000
    7.0000     5.0000
    8.0000   -13.8905
    9.0000    20.0000

forces in elements
```

EF =

```
    -5.0000     5.0000
    -0.8876     0.8876
    -0.6657    24.3343
     5.0000    -5.0000
     0.8876    -0.8876
   -24.3343     0.6657
```

9.3 A second grid example

The second grid problem is illustrated in figure 9.3. The grid is built from three elements, as illustrated. Given $E = 210$ GPa, $G = 84$ GPa, $I = 20 \times 10^{-5}$ m^4, $J = 5 \times 10^{-5}$ m^4, the MATLAB problem15.m computes displacements, reactions and stresses.

```
%............................................................................

% MATLAB codes for Finite Element Analysis
% problem15.m
% antonio ferreira 2008
```

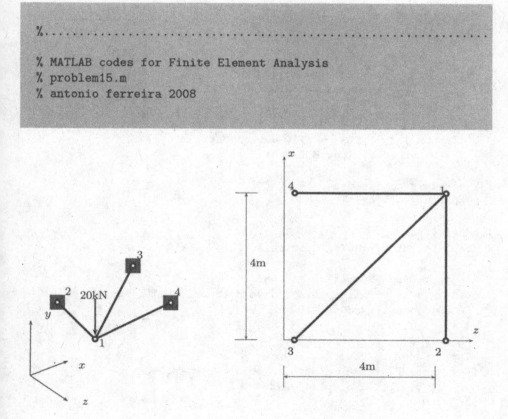

Fig. 9.3 A second grid example, problem15.m

```
% clear memory
clear all

% E; modulus of elasticity
% I: second moments of area
% J: polar moment of inertia
% G: shear modulus
% L: length of bar
E=210e6; G=84e6;  I=20e-5;   J=5e-5;

% generation of coordinates and connectivities
nodeCoordinates=[4 4; 0 4; 0 0 ; 4 0];
xx=nodeCoordinates(:,1);
yy=nodeCoordinates(:,2);
elementNodes=[1 2; 3 1; 4 1];
numberNodes=size(nodeCoordinates,1);
numberElements=size(elementNodes,1);

% GDof: global number of degrees of freedom
GDof=3*numberNodes;

force=zeros(GDof,1);
%force vector
force(1)=-20;

% computation of the system stiffness matrix
stiffness=formStiffnessGrid(GDof,numberElements,...
    elementNodes,xx,yy,E,I,G,J);

% boundary conditions
prescribedDof=[4:12]';

% solution
displacements=solution(GDof,prescribedDof,stiffness,force)

% output displacements/reactions
outputDisplacementsReactions(displacements,stiffness,...
    GDof,prescribedDof)

% % forces in elements
disp('forces in elements ')
EF=forcesInElementGrid(numberElements,elementNodes,...
    xx,yy,E,I,G,J,displacements)
```

Results for displacements, reactions and forces in elements are listed below.

Displacements

ans =

```
     1.0000    -0.0033
     2.0000     0.0010
     3.0000    -0.0010
     4.0000          0
     5.0000          0
     6.0000          0
     7.0000          0
     8.0000          0
     9.0000          0
    10.0000          0
    11.0000          0
    12.0000          0
```

reactions

ans =

```
     4.0000    10.7937
     5.0000    -1.0189
     6.0000    31.7764
     7.0000    -1.5874
     8.0000    -4.0299
     9.0000     4.0299
    10.0000    10.7937
    11.0000   -31.7764
    12.0000     1.0189
```

forces in elements

EF =

```
  -10.7937    -1.5874    10.7937
   -1.0189          0     1.0189
  -11.3984     5.6992    31.7764
   10.7937     1.5874   -10.7937
    1.0189          0    -1.0189
  -31.7764   -14.6788    11.3984
```

Chapter 10
Analysis of Timoshenko beams

10.1 Introduction

Unlike the Euler-Bernoulli beam formulation, the Timoshenko beam formulation accounts for transverse shear deformation. It is therefore capable of modeling thin or thick beams. In this chapter we perform the analysis of Timoshenko beams in static bending, free vibrations and buckling. We present the basic formulation and show how a MATLAB code can accurately solve this problem.

10.2 Formulation for static analysis

The Timoshenko theory assumes the deformed cross-section planes remain plane but not normal to the middle axis. The displacement field for this beam theory is defined as

$$u = y\theta_z, \quad w = w_0 \tag{10.1}$$

where $\theta_z(x)$ denotes the rotation of the cross-section plane about a normal to the middle axis x, and w_0 is the transverse displacement of the beam middle axis.

Normal ϵ_x and transverse shear γ_{xy} strains are defined as

$$\epsilon_x = \frac{\partial u}{\partial x} = y\frac{\partial \theta_z}{\partial x} \tag{10.2}$$

$$\gamma_{xy} = \frac{\partial u}{\partial y} + \frac{\partial w}{\partial x} = \theta_x + \frac{\partial w}{\partial x} \tag{10.3}$$

being w the transverse displacement of the beam along the y axis.

The strain energy considers both bending and shear contributions,

$$U = \frac{1}{2}\int_V \sigma_x \epsilon_x dV + \frac{1}{2}\int_V \tau_{xy}\gamma_{xy} dV \tag{10.4}$$

A.J.M. Ferreira, *MATLAB Codes for Finite Element Analysis:*
Solids and Structures, Solid Mechanics and Its Applications 157,
© Springer Science+Business Media B.V. 2009

where the normal stress is obtained by the Hooke's law as

$$\sigma_x = E\epsilon_x \tag{10.5}$$

while the transverse shear stress is obtained as

$$\tau_{xy} = kG\gamma_{xy} \tag{10.6}$$

being G the shear modulus

$$G = \frac{E}{2(1+\nu)} \tag{10.7}$$

and k the shear correction factor. This factor is dependent on the cross-section and on the type of problem. Some authors use 5/6 for static problems. Considering $dV = dAdx$ and integrating through the thickness, we obtain the strain energy in terms of the generalized displacements

$$U = \frac{1}{2}\int_V E\epsilon_x^2 dV + \frac{1}{2}\int_V kG\gamma_{xy}^2 dV =$$
$$\frac{1}{2}\int_{-a}^{a} EI_z\left(\frac{\partial\theta_z}{\partial x}\right)^2 dx + \frac{1}{2}\int_{-a}^{a} kAG\left(\frac{\partial w}{\partial x}+\theta_z\right)^2 dx \tag{10.8}$$

Each node of this two-node element considers one transverse displacement, w and one rotation θ_z, as illustrated in figure 10.1.

In opposition to Bernoulli beams, here the interpolation of displacements is independent for both w and θ_z

$$\mathbf{w} = \mathbf{N}\mathbf{w}^e \tag{10.9}$$

$$\theta_z = \mathbf{N}\theta_z^e \tag{10.10}$$

where shape functions are defined as

$$\mathbf{N} = \left[\tfrac{1}{2}(1-\xi)\quad \tfrac{1}{2}(1+\xi)\right] \tag{10.11}$$

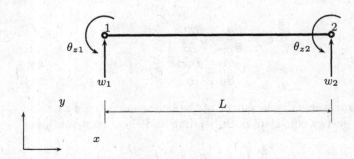

Fig. 10.1 Timoshenko beam element: degrees of freedom of the two-noded element

in natural coordinates $\xi \in [-1, +1]$. We can now compute the stiffness matrix as

$$\mathbf{K}^e = \int_{-1}^{1} \frac{EI_z}{a^2} \left(\frac{dN}{d\xi} \right)^T \left(\frac{dN}{d\xi} \right) a d\xi + \int_{-1}^{1} kGA \left(\frac{1}{a} \frac{dN}{d\xi} + N \right)^T \left(\frac{1}{a} \frac{dN}{d\xi} + N \right) a d\xi$$

(10.12)

Integrals will be computed by Gauss quadrature. Note that the bending stiffness is computed by 2×2 Gauss points, while the shear stiffness is computed with 1 Gauss point. This selective quadrature was found to be one possible solution for shear locking in thin beams [5, 12, 13].

Code **problem16.m** compute the displacements of Timoshenko beams in bending.

```
%.............................................................

% MATLAB codes for Finite Element Analysis
% problem16.m
% Timoshenko beam in bending
% antonio ferreira 2008

% clear memory
clear all

% E; modulus of elasticity
% G; shear modulus
% I: second moments of area
% L: length of beam
% thickness: thickness of beam
E=10e7; poisson = 0.30;L  = 1;thickness=0.001;
I=thickness^3/12;
EI=E*I;
kapa=5/6;
%

P = -1; % uniform pressure
% constitutive matrix
G=E/2/(1+poisson);
C=[   EI   0; 0     kapa*thickness*G];

% mesh
numberElements      = 100;
nodeCoordinates=linspace(0,L,numberElements+1);
xx=nodeCoordinates';
for i=1:size(nodeCoordinates,2)-1
```

```
     elementNodes(i,1)=i;
     elementNodes(i,2)=i+1;
end
% generation of coordinates and connectivities
numberNodes=size(xx,1);

% GDof: global number of degrees of freedom
GDof=2*numberNodes;

% computation of the system stiffness matrix
[stiffness,force]=...
     formStiffnessMassTimoshenkoBeam(GDof,numberElements,...
     elementNodes,numberNodes,xx,C,P,1,I,thickness);

% boundary conditions (simply-supported at both bords)
%fixedNodeW =[1 ; numberNodes];
%fixedNodeTX=[];
% boundary conditions (clamped at both bords)
fixedNodeW =[1 ; numberNodes];
fixedNodeTX=fixedNodeW;
% boundary conditions (cantilever)
fixedNodeW =[1];
fixedNodeTX=[1];;
prescribedDof=[fixedNodeW; fixedNodeTX+numberNodes];

% solution
displacements=solution(GDof,prescribedDof,stiffness,force);

% output displacements/reactions
outputDisplacementsReactions(displacements,stiffness,...
     GDof,prescribedDof)

U=displacements;
ws=1:numberNodes;

% max displacement
disp(' max displacement')
min(U(ws))
```

The code calls one function formStiffnessMassTimoshenkoBeam.m which computes the stiffness matrix and the force vector of the two-node Timoshenko beam (it includes the computation of the mass matrix, relevant for free vibrations, to be discussed later in this chapter).

```
function [stiffness,force,mass]=...
    formStiffnessMassTimoshenkoBeam(GDof,numberElements,...
    elementNodes,numberNodes,xx,C,P,rho,I,thickness)

% computation of stiffness matrix and force vector
% for Timoshenko beam element
stiffness=zeros(GDof);
mass=zeros(GDof);
force=zeros(GDof,1);

% stiffness matrix
    gaussLocations=[0.577350269189626;-0.577350269189626];
    gaussWeights=ones(2,1);

% bending contribution for stiffness matrix
for e=1:numberElements
    indice=elementNodes(e,:);
    elementDof=[ indice indice+numberNodes];
    indiceMass=indice+numberNodes;

    ndof=length(indice);
    length_element=xx(indice(2))-xx(indice(1));
    detJacobian=length_element/2;invJacobian=1/detJacobian;
  for q=1:size(gaussWeights,1) ;
    pt=gaussLocations(q,:);
    [shape,naturalDerivatives]=shapeFunctionL2(pt(1));
    Xderivatives=naturalDerivatives*invJacobian;
% B matrix
B=zeros(2,2*ndof);
B(1,ndof+1:2*ndof)  = Xderivatives(:)';
% K
stiffness(elementDof,elementDof)=...
    stiffness(elementDof,elementDof)+...
    B'*B*gaussWeights(q)*detJacobian*C(1,1);
force(indice)=force(indice)+...
    shape*P*detJacobian*gaussWeights(q);

mass(indiceMass,indiceMass)=mass(indiceMass,indiceMass)+...
    shape*shape'*gaussWeights(q)*I*rho*detJacobian;
mass(indice,indice)=mass(indice,indice)+shape*shape'*...
```

```
        gaussWeights(q)*thickness*rho*detJacobian;

    end
end
% shear contribution for stiffness matrix

    gaussLocations=[0.];
    gaussWeights=[2.];

for e=1:numberElements
    indice=elementNodes(e,:);
    elementDof=[ indice indice+numberNodes];
ndof=length(indice);
length_element=xx(indice(2))-xx(indice(1));
detJ0=length_element/2;invJ0=1/detJ0;
    for q=1:size(gaussWeights,1) ;
       pt=gaussLocations(q,:);
       [shape,naturalDerivatives]=shapeFunctionL2(pt(1));
       Xderivatives=naturalDerivatives*invJacobian;
% B
    B=zeros(2,2*ndof);
    B(2,1:ndof)         = Xderivatives(:)';
    B(2,ndof+1:2*ndof)  = shape;
% K
    stiffness(elementDof,elementDof)=...
         stiffness(elementDof,elementDof)+...
         B'*B*gaussWeights(q)*detJacobian*C(2,2);
   end
end
```

Distributed load P is uniform. The code is ready for simply-supported or clamped conditions at both ends. The user can easily introduce new essential boundary conditions. For simply-supported thin beams (as illustrated in figure 10.2) the analytical solution is

$$w_{max} = \frac{5PL^4}{384EI} \qquad (10.13)$$

being E the modulus of elasticity and I the second moment of area.

Displacement errors for simply-supported or clamped thin beams are below 0.1%. The model is able to model both thick or thin beams. The deformed shape for clamped-clamped beam is illustrated in figure 10.3.

The present code is also compared with exact solutions based on assumed first order shear deformation theory [14]. The analytical solution for simply-supported (SS) Timoshenko beam is

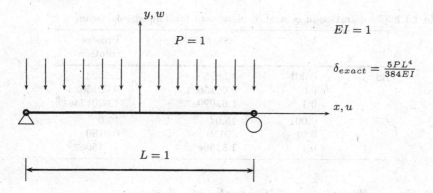

Fig. 10.2 Simply-supported Timoshenko problem, under uniform load, problem16.m

Fig. 10.3 Deformed
shape of clamped-clamped
Timoshenko beam in
bending

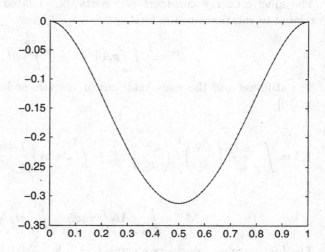

$$w(x) = \frac{PL^4}{24D}\left(\frac{x}{L} - \frac{2x^3}{L^3} + \frac{x^4}{L^4}\right) + \frac{PL^2}{2S}\left(\frac{x}{L} - \frac{x^3}{L^3}\right) \qquad (10.14)$$

being $S = kGA$ the shear stiffness, and $D = \frac{Eh^3}{12(1-\nu^2)}$ the flexural stiffness.

The analytical solution for cantilever (CF) Timoshenko beam is

$$w(x) = \frac{PL^4}{24D}\left(6\frac{x}{L} - \frac{4x^3}{L^3} + \frac{x^4}{L^4}\right) + \frac{PL^2}{2S}\left(2\frac{x}{L} - \frac{x^2}{L^2}\right) \qquad (10.15)$$

In table 10.1 we compare the present solution obtained by MATLAB code and
the exact solution by previous equations [14], for the central displacement. We
consider 100 elements and analyse various h/L ratios.

Timoshenko codes also call function **shapeFunctionL2.m** which computes shape
functions and derivatives with respect to ξ, see Section 3.3.

Table 10.1 Comparison of central displacement for Timoshenko beam

	h/L	Exact [14]	Present solution
SS	0.001	1.5625	1.5623
	0.01	0.0015631	0.0015626
	0.1	$1.62099e^{-6}$	$1.620124e^{-6}$
CF	0.001	15.0	15.0
	0.01	00150	0.0150
	0.1	$1.5156e^{-5}$	$1.5156e^{-5}$

10.3 Free vibrations of Timoshenko beams

The kinetic energy considers two parts, one related with translations and one related to rotations, in the form

$$T = \frac{1}{2} \int_{-a}^{a} \rho A \dot{w}^2 dx + \frac{1}{2} \int_{-a}^{a} \rho I_z \dot{\theta}_z^2 dx \qquad (10.16)$$

The stiffness and the mass matrices of the two-noded element can be obtained as [15]

$$\mathbf{K}^e = \int_{-1}^{1} \frac{EI_z}{a^2} \left(\frac{dN}{d\xi}\right)^T \left(\frac{dN}{d\xi}\right) a d\xi + \int_{-1}^{1} kGA \left(\frac{1}{a}\frac{dN}{d\xi} + N\right)^T \left(\frac{1}{a}\frac{dN}{d\xi} + N\right) a d\xi \qquad (10.17)$$

$$\mathbf{M}^e = \int_{-1}^{1} \rho A N^T N a d\xi + \int_{-1}^{1} \rho I_z N^T N a d\xi \qquad (10.18)$$

The first problem considers a thin ($L = 1, h = 0.001$) cantilever beam. The non-dimensional natural frequencies are given by

$$\bar{\omega} = \omega L^2 \sqrt{\frac{\rho A}{EI_z}} \qquad (10.19)$$

Results for this cantilever thin beam are presented in table 10.2. Results are in excellent agreement with exact solution [15].

```
%..............................................................

% MATLAB codes for Finite Element Analysis
% problem16vibrations.m
% Timoshenko beam in free vibrations
% antonio ferreira 2008
```

Table 10.2 Comparing natural frequencies for cantilever isotropic thin beam, using code problem16vibrations.m

Mode	Present FEM solution					Exact solution [15]
	1 elem.	2 elem.	5 elem.	10 elem.	50 elem.	
1	3.4639	3.5915	3.5321	3.5200	3.5159	3.516
2	588,390	40.3495	24.2972	22.5703	22.0439	22.035

```
% clear memory
clear all

% E; modulus of elasticity
% G; shear modulus
% I: second moments of area
% L: length of beam
% thickness: thickness of beam
E=10e7; poisson = 0.30;L  = 1;thickness=0.001;
I=thickness^3/12;
EI=E*I;
kapa=5/6;
rho=1;
A=1*thickness;
%

P = -1; % uniform pressure
% constitutive matrix
G=E/2/(1+poisson);
C=[  EI   0; 0    kapa*thickness*G];

% mesh
numberElements      = 40;
nodeCoordinates=linspace(0,L,numberElements+1);
xx=nodeCoordinates';x=xx';
for i=1:size(nodeCoordinates,2)-1
    elementNodes(i,1)=i;
    elementNodes(i,2)=i+1
end
% generation of coordinates and connectivities
```

```
numberNodes=size(xx,1);

% GDof: global number of degrees of freedom
GDof=2*numberNodes;

% computation of the system stiffness, force, mass
[stiffness,force,mass]=...
    formStiffnessMassTimoshenkoBeam(GDof,numberElements,...
    elementNodes,numberNodes,xx,C,P,rho,I,thickness);

% boundary conditions (simply-supported at both bords)
%fixedNodeW =[1 ; numberNodes];
%fixedNodeTX=[];
% boundary conditions (clamped at both bords)
fixedNodeW =[1 ; numberNodes];
fixedNodeTX=fixedNodeW;
% boundary conditions (cantilever)
fixedNodeW =[1];
fixedNodeTX=[1];;
prescribedDof=[fixedNodeW; fixedNodeTX+numberNodes];

% solution
displacements=solution(GDof,prescribedDof,stiffness,force);

% output displacements/reactions
outputDisplacementsReactions(displacements,stiffness,...
    GDof,prescribedDof)

% free vibration problem
activeDof=setdiff([1:GDof]',[prescribedDof]);
[V,D]=eig(stiffness(activeDof,activeDof),...
    mass(activeDof,activeDof));

D = diag(sqrt(D)*L*L*sqrt(rho*A/E/I));
[D,ii] = sort(D);

V1=zeros(GDof,1);
V1(activeDof,1:modeNumber)=V(:,1:modeNumber);

% drawing eigenmodes
drawEigenmodes1D(modeNumber,numberNodes,V1,xx,x)
```

Figure 10.4 illustrates the first four modes of vibration for this beam, as computed by code problem16vibrations.m, using 40 elements. This code calls function formStiffnessMassTimoshenko.m, already presented in this chapter. The code calls function drawEigenmodes1D.p which draw eigenmodes for this case.

The next example computes natural frequencies of a system suggested by Lee and Schultz [16]. We consider beams clamped or simply-supported at the ends. The non-dimensional frequencies are the same as in (10.19). The shear correction factor is taken as 5/6.

Results are listed in tables 10.3 and 10.4, and show excellent agreement with those of Lee and Schultz. Figures 10.5 and 10.6 illustrate the modes of vibration for beams clamped or simply-supported at both ends, using 40 two-noded elements.

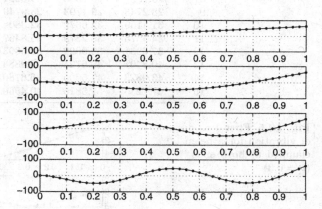

Fig. 10.4 First 4 modes of vibration for clamped beam at $x = 0$, free at $x = 1$ (with $\nu = 0.3$)

Table 10.3 Non-dimensional natural frequencies $\bar{\omega}$ for a Timoshenko beam clamped at both ends ($\nu = 0.3$, $\alpha = 5/6$, number of elements: $N = 40$)

Mode	Classical theory	h/L		
		0.002	0.01	0.1
1	4.73004	4.7345	4.7330	4.5835
2	7.8532	7.8736	7.8675	7.3468
3	10.9956	11.0504	11.0351	9.8924
4	14.1372	14.2526	14.2218	12.2118
5	17.2788	17.4888	17.4342	14.3386
6	20.4204	20.7670	20.6783	16.3046
7	23.5619	24.0955	23.9600	18.1375
8	26.7035	27.4833	27.2857	19.8593
9	29.8451	30.9398	30.6616	21.4875
10	32.9867	34.4748	34.0944	23.0358
11	36.1283	38.0993	37.5907	24.5141
12	39.2699	41.8249	41.1574	25.9179
13	42.4115	45.6642	44.8016	26.2929
14	45.5531	49.6312	48.5306	26.8419
15	48.6947	53.7410	52.3517	27.3449

Table 10.4 Non-dimensional natural frequencies $\bar{\omega}$ for a Timoshenko beam simply-supported at both ends ($\nu = 0.3$, $\alpha = 5/6$, number of elements: $N = 40$)

Mode	Classical theory	h/L		
		0.002	0.01	0.1
1	3.14159	3.1428	3.1425	3.1169
2	6.28319	6.2928	6.2908	6.0993
3	9.42478	9.4573	9.4503	8.8668
4	12.5664	12.6437	12.6271	11.3984
5	15.7080	15.8596	15.8267	13.7089
6	18.8496	19.1127	19.0552	15.8266
7	21.9911	22.4113	22.3186	17.7811
8	25.1327	25.7638	25.6231	19.5991
9	28.2743	29.1793	28.9749	21.3030
10	31.4159	32.6672	32.3806	22.9117
11	34.5575	36.2379	35.8467	24.4404
12	37.6991	39.9022	39.3803	25.9017
13	40.8407	43.6721	42.9883	26.0647
14	43.9823	47.5605	46.6780	26.2782
15	47.1239	51.5816	50.4566	26.8779

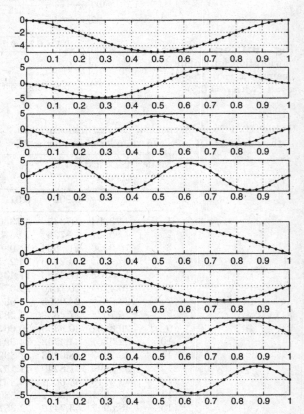

Fig. 10.5 First 4 modes of vibration for a beam clamped at both ends ($\nu = 0.3$)

Fig. 10.6 First 4 modes of vibration for a beam simply-supported at both ends ($\nu = 0.3$)

Code (problem16vibrationsSchultz.m) considers a number of boundary conditions the user should change according to the problem.

```
%..............................................................

% MATLAB codes for Finite Element Analysis
% problem16vibrationsSchultz.m
% Timoshenko beam in free vibrations
% Lee/Schultz problem
% antonio ferreira 2008

% clear memory
clear all

% E; modulus of elasticity
% G; shear modulus
% I: second moments of area
% L: length of beam
% thickness: thickness of beam
E=10e7; poisson = 0.30;L  = 1;thickness=0.1;
I=thickness^3/12;
EI=E*I;
kapa=5/6;
rho=1;
A=1*thickness;
%

P = -1; % uniform pressure
% constitutive matrix
G=E/2/(1+poisson);
C=[   EI   0; 0    kapa*thickness*G];

% mesh
numberElements      = 40;
nodeCoordinates=linspace(0,L,numberElements+1);
xx=nodeCoordinates';x=xx';
for i=1:size(nodeCoordinates,2)-1
    elementNodes(i,1)=i;
    elementNodes(i,2)=i+1
end
% generation of coordinates and connectivities
numberNodes=size(xx,1);
```

```
% GDof: global number of degrees of freedom
GDof=2*numberNodes;

% computation of the system stiffness, force, mass
[stiffness,force,mass]=...
    formStiffnessMassTimoshenkoBeam(GDof,numberElements,...
    elementNodes,numberNodes,xx,C,P,rho,I,thickness);

% boundary conditions (CC)
fixedNodeW =find(xx==min(nodeCoordinates(:))...
    | xx==max(nodeCoordinates(:)));
fixedNodeTX=fixedNodeW;
prescribedDof=[fixedNodeW; fixedNodeTX+numberNodes];

% solution
displacements=solution(GDof,prescribedDof,stiffness,force);

% output displacements/reactions
outputDisplacementsReactions(displacements,stiffness,...
    GDof,prescribedDof)

% free vibration problem
activeDof=setdiff([1:GDof]',[prescribedDof]);
[V,D]=eig(stiffness(activeDof,activeDof),...
    mass(activeDof,activeDof));
D = diag(sqrt(D)*L*L*sqrt(rho*thickness/E/I));D=sqrt(D) ;
[D,ii] = sort(D);


% lee,schultz paper
modeNumber=4;
V1=zeros(GDof,1);
V1(activeDof,1:modeNumber)=V(:,1:modeNumber);

% drawing eigenmodes
drawEigenmodes1D(modeNumber,numberNodes,V1,xx,x)
```

10.4 Buckling analysis of Timoshenko beams

The buckling analysis of Timoshenko beams considers the solution of the eigen-problem

$$[\mathbf{K} - \lambda \mathbf{K_g}] \, \mathbf{X} = 0 \qquad (10.20)$$

where λ are the critical loads and $\mathbf{X}$ the buckling modes. The geometric stiffness matrix $\mathbf{K_g}$ is obtained as [15]

$$\mathbf{K_g} = \int_0^L \left[\frac{dN_w}{dx} \right]^T P \left[\frac{dN_w}{dx} \right] dx \qquad (10.21)$$

We now consider pinned-pinned and fixed-fixed columns. The exact solution [17] is

$$P_{cr} = \frac{\pi^2 EI}{L_{eff}^2} \left[\frac{1}{1 + \frac{\pi^2 EI}{L_{eff}^2 kGA}} \right] \qquad (10.22)$$

where L_{eff} is the effective beam length. For pinned-pinned beams ($L_{eff} = L$) and for fixed-fixed beams ($L_{eff} = L/2$).

Table 10.5 shows the buckling loads for pinned-pinned and fixed-fixed Timoshenko columns. Results are in excellent agreement with those of Bazant and Cedolin [17].

Code problem16Buckling.m is listed below and calls function formStiffnessBucklingTimoshenkoBeam.m to compute the stiffness matrix and.the geometric stiffness matrix.

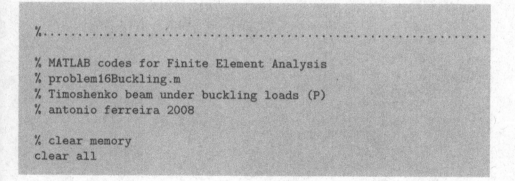

```
%.............................................................

% MATLAB codes for Finite Element Analysis
% problem16Buckling.m
% Timoshenko beam under buckling loads (P)
% antonio ferreira 2008

% clear memory
clear all
```

Table 10.5 Critical loads using 40 elements

	Pinned-pinned		Fixed-fixed	
L/h	Exact [17]	Present model	Exact [17]	Present model
10	8013.8	8021.8	29766	29877
100	8.223	8.231	32.864	32.999
1,000	0.0082	0.0082	0.0329	0.0330

```
% E; modulus of elasticity
% G; shear modulus
% I: second moments of area
% L: length of beam
% thickness: thickness of beam
E=10e6; poisson = 0.333;L  = 1;thickness=0.001;
I=thickness^3/12;
EI=E*I;
kapa=5/6;
A=1*thickness;
%

P = 1; % uniform pressure
% constitutive matrix
G=E/2/(1+poisson);
C=[   EI   0; 0    kapa*thickness*G];

% mesh
numberElements      = 40;
nodeCoordinates=linspace(0,L,numberElements+1);
xx=nodeCoordinates';x=xx';
for i=1:size(nodeCoordinates,2)-1
    elementNodes(i,1)=i;
    elementNodes(i,2)=i+1
end
% generation of coordinates and connectivities
numberNodes=size(xx,1);

% GDof: global number of degrees of freedom
GDof=2*numberNodes;

% computation of the system stiffness, Kg
[stiffness,Kg]=...
    formStiffnessBucklingTimoshenkoBeam(GDof,numberElements,...
    elementNodes,numberNodes,xx,C,P,I,thickness);

% boundary conditions (CC)
fixedNodeW =find(xx==min(nodeCoordinates(:))...
    | xx==max(nodeCoordinates(:)));
fixedNodeTX=fixedNodeW;
prescribedDof=[fixedNodeW; fixedNodeTX+numberNodes];
```

```
% boundary conditions (SS)
% fixedNodeW =find(xx==min(nodeCoordinates(:)))...
%     | xx==max(nodeCoordinates(:)));
% fixedNodeTX=[];
% prescribedDof=[fixedNodeW; fixedNodeTX+numberNodes];

% Buckling problem

    activeDof=setdiff([1:GDof]',[prescribedDof]);
    [V,D]=eig(stiffness(activeDof,activeDof),Kg(activeDof,
        activeDof));
    D=diag(D);[D,ii] = sort(D);   V = V(:,ii);

kapa=5/6;
PcrSS=pi*pi*E*I/L^2*(1/(1+pi*pi*E*I/(L*L*kapa*G*A)))
PcrCC=pi*pi*E*I/(L/2)^2*(1/(1+pi*pi*E*I/(L*L/4*kapa*G*A)))

    modeNumber=4;

    V1=zeros(GDof,1);
    V1(activeDof,1:modeNumber)=V(:,1:modeNumber);

% drawing eigenmodes
drawEigenmodes1D(modeNumber,numberNodes,V1,xx,x)
```

Code formStiffnessBucklingTimoshenkoBeam.m follows next.

```
%. . . . . . . . . . . . . . . . . . . . . . . . . . . . . . . . . . . . . . . . . . . . . . . . . . . . . . .

function [stiffness,Kg]=...
    formStiffnessBucklingTimoshenkoBeam(GDof,numberElements,...
    elementNodes,numberNodes,xx,C,P,I,thickness)

% computation of stiffness matrix and geometric stiffness
% for Timoshenko beam element
stiffness=zeros(GDof);
Kg=zeros(GDof);

% stiffness matrix
    gaussLocations=[0.];
    gaussWeights=[1.];
```

```
% bending contribution for stiffness matrix
for e=1:numberElements
    indice=elementNodes(e,:);
    elementDof=[ indice indice+numberNodes];
    ndof=length(indice);
    length_element=xx(indice(2))-xx(indice(1));
    detJacobian=length_element/2;invJacobian=1/detJacobian;
  for q=1:size(gaussWeights,1) ;
     pt=gaussLocations(q,:);
     [shape,naturalDerivatives]=shapeFunctionL2(pt(1));
     Xderivatives=naturalDerivatives*invJacobian;
% B matrix
B=zeros(2,2*ndof);
B(1,ndof+1:2*ndof)  = Xderivatives(:)';
% K
stiffness(elementDof,elementDof)=...
    stiffness(elementDof,elementDof)+...
    B'*B*gaussWeights(q)*detJacobian*C(1,1);

Kg(indice,indice)=Kg(indice,indice)+...
    Xderivatives*Xderivatives'*gaussWeights(q)*detJacobian*P;
  end
end
% shear contribution for stiffness matrix
    gaussLocations=[0.];
    gaussWeights=[2.];

for e=1:numberElements
    indice=elementNodes(e,:);
    elementDof=[ indice indice+numberNodes];
ndof=length(indice);
length_element=xx(indice(2))-xx(indice(1));
detJ0=length_element/2;invJ0=1/detJ0;
  for q=1:size(gaussWeights,1) ;
     pt=gaussLocations(q,:);
     [shape,naturalDerivatives]=shapeFunctionL2(pt(1));
     Xderivatives=naturalDerivatives*invJacobian;
% B
    B=zeros(2,2*ndof);
    B(2,1:ndof)        = Xderivatives(:)';
    B(2,ndof+1:2*ndof)  = shape;
% K
```

```
      stiffness(elementDof,elementDof)=...
          stiffness(elementDof,elementDof)+...
          B'*B*gaussWeights(q)*detJacobian*C(2,2);
   end
end
```

Figures 10.7 and 10.8 illustrate the four first buckling loads for pinned-pinned and fixed-fixed Timoshenko columns, respectively. Both beams consider $L/h = 10$, and $\nu = 0.3$.

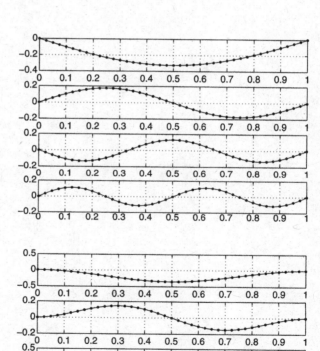

Fig. 10.7 First 4 modes of buckling for pinned-pinned Timoshenko column

Fig. 10.8 First 4 modes of buckling for fixed-fixed Timoshenko column

Chapter 11
Plane stress

11.1 Introduction

This chapter deals with the static analysis of 2D solids, particularly in plane stress. Plane stress analysis refers to problems where the thickness is quite small when compared to other dimensions in the reference plane $x-y$. The loads and boundary conditions are applied at the reference or middle plane of the structure. Displacements are computed at the reference plane. The stresses related with z coordinates are assumed to be very small and not considered in the formulation. In this chapter we consider only isotropic, homogeneous materials, and the four-node quadrilateral (Q4).

The problem is defined in a domain Ω bounded by Γ, as illustrated in figure 11.1.

11.2 Displacements, strains and stresses

The plane stress problem considers two global displacements, u and v, defined in global directions x and y, respectively.

$$\mathbf{u}(x,y) = \begin{bmatrix} u(x,y) \\ v(x,y) \end{bmatrix} \tag{11.1}$$

Strains are obtained by derivation of displacements

$$\epsilon(x,y) = \begin{bmatrix} \epsilon_x \\ \epsilon_y \\ \gamma_{xy} \end{bmatrix} = \begin{bmatrix} \dfrac{\partial u}{\partial x} \\[2mm] \dfrac{\partial v}{\partial y} \\[2mm] \dfrac{\partial u}{\partial y} + \dfrac{\partial v}{\partial x} \end{bmatrix} \tag{11.2}$$

A.J.M. Ferreira, *MATLAB Codes for Finite Element Analysis: Solids and Structures*, Solid Mechanics and Its Applications 157,
© Springer Science+Business Media B.V. 2009

Fig. 11.1 Plane stress: illustration
of domain (Ω) and boundary (Γ)

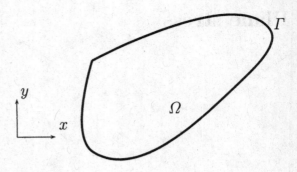

By assuming a linear elastic material, we obtain stresses as

$$\boldsymbol{\sigma} = \begin{bmatrix} \sigma_x \\ \sigma_y \\ \tau_{xy} \end{bmatrix} = \mathbf{C}\boldsymbol{\epsilon} = \begin{bmatrix} \dfrac{E}{1-\nu^2} & \dfrac{\nu E}{1-\nu^2} & 0 \\[3mm] \dfrac{\nu E}{1-\nu^2} & \dfrac{E}{1-\nu^2} & 0 \\[3mm] 0 & 0 & G = \dfrac{E}{2(1+\nu)} \end{bmatrix} \begin{bmatrix} \epsilon_x \\ \epsilon_y \\ \gamma_{xy} \end{bmatrix} \tag{11.3}$$

where E is the modulus of elasticity and ν the Poisson's coefficient.

The static equilibrium equations are defined as

$$\frac{\partial \sigma_x}{\partial x} + \frac{\partial \tau_{xy}}{\partial y} + b_x = 0 \tag{11.4}$$

$$\frac{\partial \tau_{xy}}{\partial x} + \frac{\partial \sigma_y}{\partial y} + b_y = 0 \tag{11.5}$$

where b_x, b_y are body forces.

11.3 Boundary conditions

Essential or displacement boundary conditions are applied on the boundary displacement part Γ_u, as

$$\mathbf{u} = \hat{\mathbf{u}} \tag{11.6}$$

Natural or force boundary conditions are applied on Γ_t, so that

$$\boldsymbol{\sigma}_n = \hat{\mathbf{t}} \tag{11.7}$$

where $\hat{\mathbf{t}}$ is the surface traction per unit area, and $\boldsymbol{\sigma}_n$ the normal vector to the plate.

If necessary, σ_n can be computed in natural coordinates by

$$\sigma = \begin{bmatrix} \sigma_x n_x + \tau_{xy} n_y \\ \tau_{xy} n_x + \sigma_y n_y \end{bmatrix} = \begin{bmatrix} n_x & 0 & n_y \\ 0 & n_y & n_x \end{bmatrix} \begin{bmatrix} \sigma_x \\ \sigma_y \\ \tau_{xy} \end{bmatrix} \tag{11.8}$$

11.4 Potential energy

The potential energy can be defined as

$$\Pi = U - W \tag{11.9}$$

where U is the elastic strain deformation,

$$U = \frac{1}{2} \int_{\Omega} h \epsilon^T \sigma d\Omega = \frac{1}{2} \int_{\Omega} h \epsilon^T \mathbf{C} \epsilon d\Omega \tag{11.10}$$

The energy produced by the external forces is given by

$$W = \int_{\Omega} h \mathbf{u}^T \mathbf{b} d\Omega + \int_{\Gamma_t} h \mathbf{u}^T \hat{\mathbf{t}} d\Gamma \tag{11.11}$$

11.5 Finite element discretization

Given a domain denoted by Ω^e and a boundary denoted by Γ^e, the n-noded finite element displacement vector is defined by $2n$ degrees of freedom,

$$\mathbf{u}^e = \begin{bmatrix} u_1 & v_1 & u_2 & v_2 & \dots & u_n & v_n \end{bmatrix}^T \tag{11.12}$$

11.6 Interpolation of displacements

The displacement vector in each element is interpolated by the nodal displacements as

$$u = \sum_{i=1}^{n} N_i^e u_i; \quad v = \sum_{i=1}^{n} N_i^e v_i \tag{11.13}$$

where N_i^e denote the element shape functions. This can also be expressed in matrix form as

$$\mathbf{u} = \begin{bmatrix} N_1^e & 0 & 0 & N_2^e & 0 & ... & N_n^e & 0 \\ 0 & N_1^e & 0 & 0 & N_2^e & 0 & ... & N_n^e \end{bmatrix} \mathbf{u}^e = \mathbf{N}\mathbf{u}^e \tag{11.14}$$

The strain vector can be obtained by derivation of the displacements as

$$\boldsymbol{\epsilon} = \begin{bmatrix} \dfrac{\partial N_1^e}{\partial x} & 0 & \dfrac{\partial N_2^e}{\partial x} & 0 & ... & \dfrac{\partial N_n^e}{\partial x} & 0 \\[2mm] 0 & \dfrac{\partial N_1^e}{\partial y} & 0 & \dfrac{\partial N_2^e}{\partial y} & 0 & ... & \dfrac{\partial N_n^e}{\partial y} \\[2mm] \dfrac{\partial N_1^e}{\partial y} & \dfrac{\partial N_1^e}{\partial x} & \dfrac{\partial N_2^e}{\partial y} & \dfrac{\partial N_2^e}{\partial x} & ... & \dfrac{\partial N_n^e}{\partial y} & \dfrac{\partial N_n^e}{\partial x} \end{bmatrix} \mathbf{u}^e = \mathbf{B}\mathbf{u}^e \tag{11.15}$$

where $\mathbf{B}$ is the strain-displacement matrix. This matrix is needed for computation of the stiffness matrix, and the stress vector at each element.

11.7 Element energy

The total potential energy can de defined at each element by

$$\Pi^e = U^e - W^e \tag{11.16}$$

where the strain energy is defined as

$$U^e = \frac{1}{2} \int_{\Omega^e} h\boldsymbol{\epsilon}^T \boldsymbol{\sigma} d\Omega^e = \frac{1}{2} \int_{\Omega^e} h\boldsymbol{\epsilon}^T \mathbf{C}\boldsymbol{\epsilon} d\Omega^e \tag{11.17}$$

and the energy produced by the surface tractions is given by

$$W^e = \int_{\Omega^e} h\mathbf{u}^T \mathbf{b} d\Omega^e + \int_{\Gamma^e} h\mathbf{u}^T \hat{\mathbf{t}} d\Gamma^e \tag{11.18}$$

We can introduce these expressions into the total potential energy as

$$\Pi^e = \frac{1}{2} \mathbf{u}^{eT} \mathbf{K}^e \mathbf{u}^e - \mathbf{u}^{eT} \mathbf{f}^e \tag{11.19}$$

where the element stiffness matrix is obtained as

$$\mathbf{K}^e = \int_{\Omega^e} h\mathbf{B}^T \mathbf{C}\mathbf{B} d\Omega^e \tag{11.20}$$

and the vector of nodal forces is obtained as

$$\mathbf{f}^e = \int_{\Omega^e} h\mathbf{N}^T \mathbf{b} d\Omega^e + \int_{\Gamma^e} h\mathbf{N}^T \hat{\mathbf{t}} d\Gamma^e \tag{11.21}$$

11.8 Quadrilateral element Q4

We consider a quadrilateral element, illustrated in figure 11.2. The element is defined by four nodes in natural coordinates (ξ, η). The coordinates are interpolated as

$$x = \sum_{i=1}^{4} N_i x_i; \quad y = \sum_{i=1}^{4} N_i y_i \qquad (11.22)$$

where N_i are the Lagrange shape functions, given by

$$N_1(\xi, \eta) = l_1(\xi) l_1(\eta) = \frac{1}{4}(1 - \xi)(1 - \eta) \qquad (11.23)$$

$$N_2(\xi, \eta) = l_2(\xi) l_1(\eta) = \frac{1}{4}(1 + \xi)(1 - \eta) \qquad (11.24)$$

$$N_3(\xi, \eta) = l_2(\xi) l_2(\eta) = \frac{1}{4}(1 + \xi)(1 + \eta) \qquad (11.25)$$

$$N_4(\xi, \eta) = l_1(\xi) l_2(\eta) = \frac{1}{4}(1 - \xi)(1 + \eta) \qquad (11.26)$$

Displacements are interpolated as

$$u = \sum_{i=1}^{4} N_i u_i; \quad v = \sum_{i=1}^{4} N_i v_i \qquad (11.27)$$

where u, v are the displacements at any point in the element and $u_i, v_i; \; i = 1, ..., n$ the nodal displacements.

Derivatives $\dfrac{\partial}{\partial \xi}, \dfrac{\partial}{\partial \eta}$, can be found as

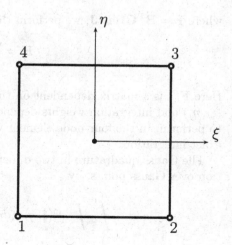

Fig. 11.2 Quadrilateral Q4 element in natural coordinates

$$\begin{bmatrix} \dfrac{\partial}{\partial \xi} \\[3mm] \dfrac{\partial}{\partial \eta} \end{bmatrix} = \begin{bmatrix} \dfrac{\partial x}{\partial \xi} & \dfrac{\partial y}{\partial \xi} \\[3mm] \dfrac{\partial x}{\partial \eta} & \dfrac{\partial y}{\partial \eta} \end{bmatrix} \begin{bmatrix} \dfrac{\partial}{\partial x} \\[3mm] \dfrac{\partial}{\partial y} \end{bmatrix} \qquad (11.28)$$

In matricial form, we can write relations (11.28) as

$$\frac{\partial}{\partial \xi} = \mathbf{J}\frac{\partial}{\partial \mathbf{x}} \qquad (11.29)$$

where $\mathbf{J}$ is the Jacobian operator, relating natural and global coordinates. The derivatives with respect to the global coordinates can be found as

$$\frac{\partial}{\partial \mathbf{x}} = \mathbf{J}^{-1}\frac{\partial}{\partial \xi} \qquad (11.30)$$

Note that in very distorted elements the Jacobian inverse, $\mathbf{J}^{-1}$ may not exist.

The stiffness matrix is then obtained by

$$\mathbf{K} = \int_V \mathbf{B}^T \mathbf{C} \mathbf{B} dV \qquad (11.31)$$

Note that $\mathbf{B}$ depends on the element natural coordinates ξ, η. The element volume is given by

$$dV = h \; det\mathbf{J} d\xi d\eta \qquad (11.32)$$

where $det\mathbf{J}$ is the determinant of the Jacobian matrix and h the thickness of the plate. The integral in the stiffness matrix is computed numerically by Gauss quadrature in two dimensions. Taking

$$\mathbf{K} = h \int_A \mathbf{F} d\xi d\eta \qquad (11.33)$$

where $\mathbf{F} = \mathbf{B}^T \mathbf{C} \mathbf{B} det\mathbf{J}$, we perform the numerical computation by

$$\mathbf{K} = \sum_{i,j,k} \mathbf{F}_{i,j} \alpha_{i,j} \qquad (11.34)$$

Here $\mathbf{F}_{i,j}$ is a matrix dependent on the natural points (ξ_i, η_j). Integration points (ξ_i, η_j) and integration weights depend on the type of integration the user wishes to perform. In the four-node element we can use a 2×2 numerical integration for exact integration.

The Gauss quadrature in two dimensions replaces the integration by a summation over Gauss points, by

$$\int_{-1}^{1}\int_{-1}^{1} F(\xi, \eta) d\xi d\eta = \sum_{i=1}^{p}\sum_{j=1}^{q} w_i w_j F(\xi_i, \eta_j) \qquad (11.35)$$

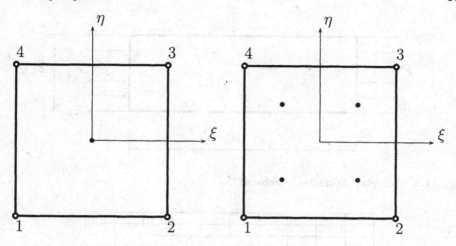

Fig. 11.3 One Gauss point integration ($\xi = 0, \eta = 0$); two Gauss point integration ($\xi, \eta = \pm\frac{1}{\sqrt{3}}$)

where p, q are the number of integrating points in the ξ, η directions, respectively, and w_i, w_j are the corresponding weights, as illustrated in figure 11.3, for some integration orders.

The stiffness matrix can be computed using 2×2 Gauss points as

$$\mathbf{K}^e = \int_{\Omega^e} h\mathbf{B}^T\mathbf{C}\mathbf{B}d\Omega^e = \int_{-1}^{1}\int_{-1}^{1} h\mathbf{B}^T\mathbf{C}\mathbf{B}det\mathbf{J}d\xi d\eta = h\sum_{i=1}^{2}\sum_{j=1}^{2} \mathbf{B}^T\mathbf{C}\mathbf{B}det\mathbf{J}w_i w_j \tag{11.36}$$

All Gauss points have unit weight, in this integration rule.

11.9 Example: plate in traction

We consider a thin plate under uniform traction forces at its extremes. The problem is illustrated in figure 11.4.

In figure 11.5 we show the finite element mesh considering 10×5 elements. In figure 11.6 the deformed shape of the problem is illustrated.

The MATLAB code for this problem is (problem17.m).

```
%. . . . . . . . . . . . . . . . . . . . . . . . . . . . . . . . . . . . . . . . . . . . . . . . . . . . . . . . . . . . . . . . . . . . . . . . . . . . . . . . . . . . . . . . . . . . . . .

% MATLAB codes for Finite Element Analysis
```

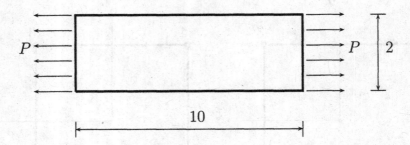

Fig. 11.4 Thin plate in traction, problem17.m

Fig. 11.5 Finite element mesh for a thin plate in tension

Displacement filed in X direction

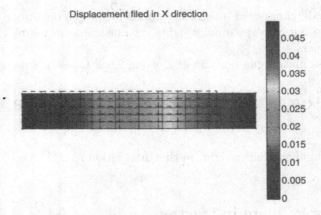

Fig. 11.6 Plate in traction: displacements in XX

```
% problem17.m
% 2D problem: thin plate in tension
% antonio ferreira 2008

% clear memory
clear all;colordef white;clf
```

```
% materials
E  = 10e7;      poisson = 0.30;

% matriz C
C=E/(1-poisson^2)*[1 poisson 0;poisson 1 0;0 0 (1-poisson)/2];

% load
P = 1e6;

%Mesh generation
Lx=5;
Ly=1;
numberElementsX=20;
numberElementsY=10;
numberElements=numberElementsX*numberElementsY;
[nodeCoordinates, elementNodes] = ...
    rectangularMesh(Lx,Ly,numberElementsX,numberElementsY);
xx=nodeCoordinates(:,1);
yy=nodeCoordinates(:,2);
drawingMesh(nodeCoordinates,elementNodes,'Q4','k-');
numberNodes=size(xx,1);

% GDof: global number of degrees of freedom
GDof=2*numberNodes;

% calculation of the system stiffness matrix
stiffness=formStiffness2D(GDof,numberElements,...
    elementNodes,numberNodes,nodeCoordinates,C,1,1);

% boundary conditions
fixedNodeX=find(nodeCoordinates(:,1)==0);  % fixed in XX
fixedNodeY=find(nodeCoordinates(:,2)==0);  % fixed in YY
prescribedDof=[fixedNodeX; fixedNodeY+numberNodes];

% force vector (distributed load applied at xx=Lx)
force=zeros(GDof,1);
rightBord=find(nodeCoordinates(:,1)==Lx);
force(rightBord)=P*Ly/numberElementsY;
force(rightBord(1))=P*Ly/numberElementsY/2;
force(rightBord(end))=P*Ly/numberElementsY/2;

% solution
displacements=solution(GDof,prescribedDof,stiffness,force);
```

```
% displacements
disp('Displacements')
jj=1:GDof; format
f=[jj; displacements'];
fprintf('node U\n')
fprintf('%3d %12.8f\n',f)
UX=displacements(1:numberNodes);
UY=displacements(numberNodes+1:GDof);
scaleFactor=10;

% deformed shape
figure
drawingField(nodeCoordinates+scaleFactor*[UX UY],...
    elementNodes,'Q4',UX);%U XX
hold on
drawingMesh(nodeCoordinates+scaleFactor*[UX UY],...
    elementNodes,'Q4','k-');
drawingMesh(nodeCoordinates,elementNodes,'Q4','k--');
colorbar
title('U XX  (on deformed shape)')
axis off

% stresses at nodes
stresses2D(GDof,numberElements,elementNodes,numberNodes,...
    nodeCoordinates,displacements,UX,UY,C,scaleFactor)
```

11.10 Example: beam in bending

We show in this example (code problem18.m) how MATLAB can help in the computation of a beam in bending (figure 11.7). Note some of the differences to problem17.m:

- The boundary conditions are different, in this case both u and v are fixed at $x = 0$.
- The applied force is in the y direction, so care must be taken to ensure that degrees of freedom are properly assigned.

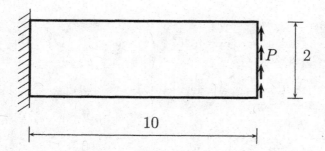

Fig. 11.7 Thin plate in bending, problem18.m

```
%.............................................................

% MATLAB codes for Finite Element Analysis
% problem18.m
% 2D problem: beam in bending
% antonio ferreira 2008

% clear memory
clear all;colordef white;clf

% materials
E = 10e7;      poisson = 0.30;

% matriz C
C=E/(1-poisson^2)*[1 poisson 0;poisson 1 0;0 0 (1-poisson)/2];

% load
P = 1e6;

%Mesh generation
Lx=5;
Ly=1;
numberElementsX=20;
numberElementsY=10;
numberElements=numberElementsX*numberElementsY;
[nodeCoordinates, elementNodes] = ...
    rectangularMesh(Lx,Ly,numberElementsX,numberElementsY);
xx=nodeCoordinates(:,1);
yy=nodeCoordinates(:,2);
```

```
drawingMesh(nodeCoordinates,elementNodes,'Q4','k-');
numberNodes=size(xx,1);

% GDof: global number of degrees of freedom
GDof=2*numberNodes;

% computation of the system stiffness matrix
stiffness=formStiffness2D(GDof,numberElements,...
    elementNodes,numberNodes,nodeCoordinates,C,1,1);

% boundary conditions
fixedNodeX=find(nodeCoordinates(:,1)==0);  % fixed in XX
fixedNodeY=find(nodeCoordinates(:,1)==0);  % fixed in YY
prescribedDof=[fixedNodeX; fixedNodeY+numberNodes];

% force vector (distributed load applied at xx=Lx)
force=zeros(GDof,1);
rightBord=find(nodeCoordinates(:,1)==Lx);
force(rightBord+numberNodes)=P*Ly/numberElementsY;
force(rightBord(1)+numberNodes)=P*Ly/numberElementsY/2;
force(rightBord(end)+numberNodes)=P*Ly/numberElementsY/2;
% solution
displacements=solution(GDof,prescribedDof,stiffness,force);

% displacements and deformed shape
disp('Displacements')
jj=1:GDof; format
f=[jj; displacements'];
fprintf('node U\n')
fprintf('%3d %12.8f\n',f)
UX=displacements(1:numberNodes);
UY=displacements(numberNodes+1:GDof);
scaleFactor=0.1;

figure
drawingField(nodeCoordinates+scaleFactor*[UX UY],...
    elementNodes,'Q4',UX);%U XX
hold on
drawingMesh(nodeCoordinates+scaleFactor*[UX UY],...
    elementNodes,'Q4','k-');
drawingMesh(nodeCoordinates,elementNodes,'Q4','k--');
colorbar
title('U XX  (on deformed shape)')
```

```
axis off

% stresses at nodes
stresses2D(GDof,numberElements,...
    elementNodes,numberNodes,nodeCoordinates,...
    displacements,UX,UY,C,scaleFactor);
```

In figure 11.8 we show the evolution of displacements u on top of the deformed shape of the beam. If the user wishes to plot another displacement, just change the number of the displacement component upon calling drawingField.m.

In figure 11.9 we show the evolution of σ_x stress in the beam. If the user wishes to plot another stress, just change the number of the stress component upon calling stresses2D.m.

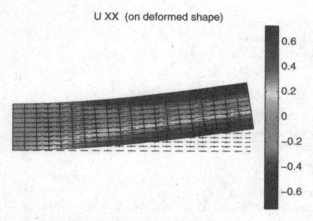

Fig. 11.8 Beam in bending: plot of displacements u

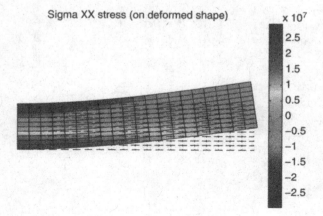

Fig. 11.9 Beam in bending: plot of σ_x stresses

Functions formStiffness2D.m and stresses2D.m are listed below.

```
%..................................................................

function [stiffness,mass]=formStiffness2D(GDof,numberElements,...
    elementNodes,numberNodes,nodeCoordinates,C,rho,thickness)

% compute stiffness matrix (and mass matrix)
% for plane stress Q4 elements

stiffness=zeros(GDof);
mass=zeros(GDof);

% 2 by 2 quadrature
[gaussWeights,gaussLocations]=gaussQuadrature('complete');

for e=1:numberElements
  indice=elementNodes(e,:);
  elementDof=[ indice indice+numberNodes ];
  ndof=length(indice);

  % cycle for Gauss point
  for q=1:size(gaussWeights,1)
    GaussPoint=gaussLocations(q,:);
    xi=GaussPoint(1);
    eta=GaussPoint(2);

% shape functions and derivatives
    [shapeFunction,naturalDerivatives]=shapeFunctionQ4(xi,eta)

% Jacobian matrix, inverse of Jacobian,
% derivatives w.r.t. x,y
    [Jacob,invJacobian,XYderivatives]=...
        Jacobian(nodeCoordinates(indice,:),naturalDerivatives);

%  B matrix
    B=zeros(3,2*ndof);
    B(1,1:ndof)        = XYderivatives(:,1)';
    B(2,ndof+1:2*ndof) = XYderivatives(:,2)';
    B(3,1:ndof)        = XYderivatives(:,2)';
    B(3,ndof+1:2*ndof) = XYderivatives(:,1)';

% stiffness matrix
```

```
        stiffness(elementDof,elementDof)=...
            stiffness(elementDof,elementDof)+...
            B'*C*thickness*B*gaussWeights(q)*det(Jacob);
% mass matrix
        mass(indice,indice)=mass(indice,indice)+...
            shapeFunction*shapeFunction'*...
            rho*thickness*gaussWeights(q)*det(Jacob);
        mass(indice+numberNodes,indice+numberNodes)=...
            mass(indice+numberNodes,indice+numberNodes)+...
            shapeFunction*shapeFunction'*...
            rho*thickness*gaussWeights(q)*det(Jacob);

    end
end
```

```
%...............................................................

function stresses2D(GDof,numberElements,...
    elementNodes,numberNodes,nodeCoordinates,...
    displacements,UX,UY,C,scaleFactor)

% 2 by 2 quadrature
[gaussWeights,gaussLocations]=gaussQuadrature('complete');

% stresses at nodes
stress=zeros(numberElements,size(elementNodes,2),3);
stressPoints=[-1 -1;1 -1;1 1;-1 1];

for e=1:numberElements
  indice=elementNodes(e,:);
  elementDof=[ indice indice+numberNodes ];
  nn=length(indice);
  for q=1:size(gaussWeights,1)
    pt=gaussLocations(q,:);
    wt=gaussWeights(q);
    xi=pt(1);
    eta=pt(2);
% shape functions and derivatives
    [shapeFunction,naturalDerivatives]=shapeFunctionQ4(xi,eta)

% Jacobian matrix, inverse of Jacobian,
```

```
% derivatives w.r.t. x,y
    [Jacob,invJacobian,XYderivatives]=...
        Jacobian(nodeCoordinates(indice,:),naturalDerivatives);

% B matrix
    B=zeros(3,2*nn);
    B(1,1:nn)       = XYderivatives(:,1)';
    B(2,nn+1:2*nn)  = XYderivatives(:,2)';
    B(3,1:nn)       = XYderivatives(:,2)';
    B(3,nn+1:2*nn)  = XYderivatives(:,1)';

% element deformation
    strain=B*displacements(elementDof);
    stress(e,q,:)=C*strain;
  end
end

% drawing stress fields
% on top of the deformed shape
figure
drawingField(nodeCoordinates+scaleFactor*[UX UY],...
    elementNodes,'Q4',stress(:,:,1));%sigma XX
hold on
drawingMesh(nodeCoordinates+scaleFactor*[UX UY],...
    elementNodes,'Q4','k-');
drawingMesh(nodeCoordinates,elementNodes,'Q4','k--');
colorbar
title('Sigma XX stress (on deformed shape)')
axis off
```

Some functions are used for all 2D problems (even for plates). Function shapeFunctionQ4.m computes the shape functions and derivatives of the shape functions with respect to natural ξ,η coordinates. Function Jacobian.m computes the Jacobian matrix, and its inverse. The computation of Gauss point locations and weights is made in function gaussQuadrature.m. The listing of these functions is given as follows

```
% ...........................................................
    function [shape,naturalDerivatives]=shapeFunctionQ4(xi,eta)

    % shape function and derivatives for Q4 elements
```

```
% shape : Shape functions
% naturalDerivatives: derivatives w.r.t. xi and eta
% xi, eta: natural coordinates (-1 ... +1)

shape=1/4*[ (1-xi)*(1-eta);(1+xi)*(1-eta);
       (1+xi)*(1+eta);(1-xi)*(1+eta)];
naturalDerivatives=...
     1/4*[-(1-eta), -(1-xi);1-eta,    -(1+xi);
     1+eta,        1+xi;-(1+eta),    1-xi];

end % end function shapeFunctionQ4
```

```
% ..............................................................

function [JacobianMatrix,invJacobian,XYDerivatives]=...
    Jacobian(nodeCoordinates,naturalDerivatives)

% JacobianMatrix     : Jacobian matrix
% invJacobian : inverse of Jacobian Matrix
% XYDerivatives  : derivatives w.r.t. x and y
% naturalDerivatives  : derivatives w.r.t. xi and eta
% nodeCoordinates  : nodal coordinates at element level

JacobianMatrix=nodeCoordinates'*naturalDerivatives;
invJacobian=inv(JacobianMatrix);
XYDerivatives=naturalDerivatives*invJacobian;

end % end function Jacobian
```

```
% ..............................................................

function [weights,locations]=gaussQuadrature(option)
% Gauss quadrature for Q4 elements
% option 'complete' (2x2)
% option 'reduced'  (1x1)
% locations: Gauss point locations
% weights: Gauss point weights
```

```
switch option
    case 'complete'

    locations=...
      [ -0.577350269189626  -0.577350269189626;
         0.577350269189626  -0.577350269189626;
         0.577350269189626   0.577350269189626;
        -0.577350269189626   0.577350269189626];
    weights=[ 1;1;1;1];

    case 'reduced'

    locations=[0 0];
    weights=[4];
  end

  end   % end function gaussQuadrature
```

Note that we analyse free vibrations with 2D elements, the function formStiff-
ness2D.m already computes the mass matrix. The analysis is left to the reader.

Chapter 12
Analysis of Mindlin plates

12.1 Introduction

This chapter considers the static analysis of Mindlin plates in bending. Also, we compute the free vibration problem and the buckling problem.

12.2 The Mindlin plate theory

The Mindlin plate theory or first-order shear deformation theory for plates includes the effect of transverse shear deformations [18]. It may be considered an extension of the Timoshenko theory for beams in bending. The main difference for thin, Kirchhoff-type theories is that in the Mindlin theory the normals to the undeformed middle plane of the plate remain straight, but not normal to the deformed middle surface.

The strain energy of the Mindlin plate is given as [15, 18]

$$U = \frac{1}{2} \int_V \sigma_f^T \epsilon_f dV + \frac{\alpha}{2} \int_V \sigma_c^T \epsilon_c dV \qquad (12.1)$$

where

$$\sigma_f^T = [\sigma_x \ \sigma_y \ \tau_{xy}] \qquad (12.2)$$

$$\epsilon_f^T = [\epsilon_x \ \epsilon_y \ \gamma_{xy}] \qquad (12.3)$$

are the bending stresses and strains, and

$$\sigma_c^T = [\tau_{xz} \ \tau_{yz}] \qquad (12.4)$$

$$\epsilon_c^T = [\gamma_{xz} \ \gamma_{yz}] \qquad (12.5)$$

are the transverse shear stresses and strains. The α parameter, also known as the shear correction factor can be taken as $5/6$ [4].

A.J.M. Ferreira, *MATLAB Codes for Finite Element Analysis:*
Solids and Structures, Solid Mechanics and Its Applications 157,
© Springer Science+Business Media B.V. 2009

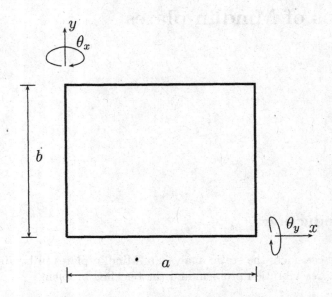

Fig. 12.1 Mindlin plate: illustration of geometry, degrees of freedom

The assumed displacement field for a thick plate (thickness h) is defined as

$$u = z\theta_x; \quad v = z\theta_y; \quad w = w_0 \tag{12.6}$$

where θ_x, θ_y are the rotations of the normal to the middle plane with respect to axes y and x, respectively (as illustrated in figure 12.1).

12.2.1 Strains

Bending strains are obtained as

$$\epsilon_x = \frac{\partial u}{\partial x} = z\frac{\partial \theta_x}{\partial x} \tag{12.7}$$

$$\epsilon_y = \frac{\partial v}{\partial y} = z\frac{\partial \theta_y}{\partial y} \tag{12.8}$$

$$\gamma_{xy} = \frac{\partial u}{\partial y} + \frac{\partial v}{\partial x} = z\left(\frac{\partial \theta_y}{\partial x} + \frac{\partial \theta_x}{\partial y}\right) \tag{12.9}$$

while the transverse shear deformations are obtained as

$$\gamma_{xz} = \frac{\partial w}{\partial x} + \frac{\partial u}{\partial z} = \frac{\partial w}{\partial x} + \theta_x \qquad (12.10)$$

$$\gamma_{yz} = \frac{\partial w}{\partial y} + \frac{\partial v}{\partial z} = \frac{\partial w}{\partial y} + \theta_y \qquad (12.11)$$

12.2.2 Stresses

The linear elastic stress-strain relations in bending are defined for a homogeneous, isotropic material as

$$\sigma_f = \mathbf{D_f}\epsilon_f \qquad (12.12)$$

where $\mathbf{D_f}$ is defined as

$$\mathbf{D_f} = \frac{E}{1-\nu^2} \begin{bmatrix} 1 & \nu & 0 \\ \nu & 1 & 0 \\ 0 & 0 & \dfrac{1-\nu}{2} \end{bmatrix} \qquad (12.13)$$

while the linear elastic stress-strain relations in transverse shear are defined as

$$\sigma_c = \mathbf{D_c}\epsilon_c \qquad (12.14)$$

where

$$\mathbf{D_c} = \begin{bmatrix} G & 0 \\ 0 & G \end{bmatrix} \qquad (12.15)$$

where G the shear modulus. Introducing these concepts into the strain energy (12.1), we obtain

$$U = \frac{1}{2}\int_V \epsilon_f^T D_f \epsilon_f dV + \frac{\alpha}{2}\int_V \epsilon_c^T \mathbf{D_c}\epsilon_c dV \qquad (12.16)$$

12.3 Finite element discretization

The generalized displacements are independently interpolated using the same shape functions

$$w = \sum_{i=1}^{n} N_i(\xi,\eta)w_i, \quad \theta_x = \sum_{i=1}^{n} N_i(\xi,\eta)\theta_{xi}, \quad \theta_y = \sum_{i=1}^{n} N_i(\xi,\eta)\theta_{yi} \qquad (12.17)$$

where $N_i(\xi,\eta)$ are the shape functions of a bilinear four-noded Q4 element.
  Strains are defined as

$$\epsilon_f = z\mathbf{B}_f \mathbf{d}^e; \quad \epsilon_c = \mathbf{B}_c \mathbf{d}^e \qquad (12.18)$$

The strain-displacement matrices for bending and shear contributions are obtained by derivation of the shape functions by

$$\mathbf{B}_f = \begin{bmatrix} 0 & \dfrac{\partial N_1}{\partial x} & 0 & \dots & 0 & \dfrac{\partial N_4}{\partial x} & 0 \\[2mm] 0 & 0 & \dfrac{\partial N_1}{\partial y} & \dots & 0 & 0 & \dfrac{\partial N_4}{\partial y} \\[2mm] 0 & \dfrac{\partial N_1}{\partial y} & \dfrac{\partial N_1}{\partial x} & \dots & 0 & \dfrac{\partial N_4}{\partial y} & \dfrac{\partial N_4}{\partial x} \end{bmatrix} \tag{12.19}$$

$$\mathbf{B}_c = \begin{bmatrix} \dfrac{\partial N_1}{\partial x} & \cdot N_1 & 0 & \dots & \dfrac{\partial N_4}{\partial x} & N_4 & 0 \\[2mm] \dfrac{\partial N_1}{\partial y} & 0 & N_1 & \dots & \dfrac{\partial N_4}{\partial y} & 0 & N_4 \end{bmatrix} \tag{12.20}$$

where

$$\mathbf{d}^{eT} = \{ w_1 \quad \theta_{x1} \quad \theta_{y1} \dots \quad w_4 \quad \theta_{x4} \quad \theta_{y4} \} \tag{12.21}$$

We then obtain the plate strain energy as

$$U = \frac{1}{2}\mathbf{d}^{eT} \int_{\Omega^e} \int_z \mathbf{B}_f^T \mathbf{D_f} \mathbf{B}_f \, dz d\Omega^e \mathbf{d^e} + \frac{\alpha}{2}\mathbf{d}^{eT} \int_{\Omega^e} \int_z \mathbf{B}_c^T \mathbf{D_c} \mathbf{B}_c \, dz d\Omega^e \mathbf{d^e} \tag{12.22}$$

The stiffness matrix of the Mindlin plate is then obtained as

$$\mathbf{K}^e = \frac{h^3}{12} \int_{\Omega^e} \mathbf{B}_f^T \mathbf{D_f} \mathbf{B}_f d\Omega^e + \alpha h \int_{\Omega^e} \mathbf{B}_c^T \mathbf{D_c} \mathbf{B}_c d\Omega^e \tag{12.23}$$

or

$$\mathbf{K}^e = \frac{h^3}{12} \int_{-1}^{1} \int_{-1}^{1} \mathbf{B}_f^T D_f \mathbf{B}_f |J| d\xi d\eta + \alpha h \int_{-1}^{1} \int_{-1}^{1} \mathbf{B}_c^T D_C \mathbf{B}_c |J| d\xi d\eta \tag{12.24}$$

where $|J|$ is the determinant of the Jacobian matrix.

The vector of nodal forces equivalent to distributed forces P is defined as

$$\mathbf{f}^e = \int_{-1}^{1} \int_{-1}^{1} \mathbf{N} \, P \, |J| d\xi d\eta \tag{12.25}$$

Both the stiffness matrix and the force vector integrals are computed by numerical integration. The stiffness integral is solved by considering for the Q4 element, 2×2 Gauss points for the bending contribution and 1 point for the shear contribution. This selective integration proved to be one of the simplest remedies for avoiding shear locking [4, 12].

12.4 Example: a square Mindlin plate in bending

We consider a simply-supported and clamped square plate (side $a = 1$) under uniform transverse pressure ($P = 1$), and thickness h. The modulus of elasticity is taken $E = 10,920$[1] and the Poisson's ratio is taken as $\nu = 0.3$. The non-dimensional transverse displacement is set as

$$\bar{w} = w \frac{D}{Pl^4} \qquad (12.26)$$

where the bending stiffness D is taken as

$$D = \frac{Eh^3}{12(1 - \nu^2)} \qquad (12.27)$$

In table 12.1 we present non-dimensional transverse displacement results obtained by the code problem19.m for various thickness values and boundary conditions. In figure 12.2 we show the deformed shape of a simply-supported plate, using a 20×20 Q4 mesh.

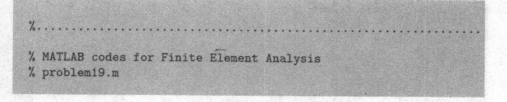

```
%......................................................................

% MATLAB codes for Finite Element Analysis
% problem19.m
```

Table 12.1 Non-dimensional transverse displacement of a square plate, under uniform pressure – simply-supported (SSSS) and clamped (CCCC) boundary conditions

a/h	Mesh	SSSS	CCCC
10	2 × 2	0.003545	0.000357
	6 × 6	0.004245	0.001486
	10 × 10	0.004263	0.001498
	20 × 20	0.004270	0.001503
	30 × 30	0.004271	0.001503
	Exact solution	0.004270	
10,000	2 × 2	0.003188	$3.5e^{-10}$
	6 × 6	0.004024	0.001239
	10 × 10	0.004049	0.001255
	20 × 20	0.004059	0.001262
	30 × 30	0.004060	0.001264
	Exact solution	0.004060	0.001260

[1] The reader may be curious about the reason for this particular value of E. With $a = 1$, thickness $h = 0.1$ and the mentioned values for E and ν we obtain a flexural stiffness of 1. This is only a practical convenience for non-dimensional results, not really a meaningful value.

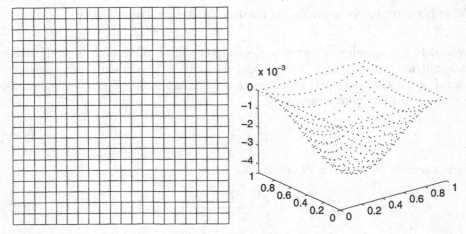

Fig. 12.2 Mesh of 20 × 20 Q4 elements and deformed shape

```
% Mindlin plate in bending
% antonio ferreira 2008

% clear memory
clear all;colordef white;clf

% materials
E  = 10920;        poisson = 0.30; kapa=5/6;
thickness=0.1;
I=thickness^3/12;

% matrix C
% bending part
C_bending=I*E/(1-poisson^2)*...
    [1 poisson 0;poisson 1 0;0 0 (1-poisson)/2];
% shear part
C_shear=kapa*thickness*E/2/(1+poisson)*eye(2);

% load
P = -1;

%Mesh generation
L  = 1;
numberElementsX=20;
numberElementsY=20;
numberElements=numberElementsX*numberElementsY;
```

```
%
[nodeCoordinates, elementNodes] = ...
    rectangularMesh(L,L,numberElementsX,numberElementsY);
xx=nodeCoordinates(:,1);
yy=nodeCoordinates(:,2);
drawingMesh(nodeCoordinates,elementNodes,'Q4','k-');
axis off
numberNodes=size(xx,1);

% GDof: global number of degrees of freedom
GDof=3*numberNodes;

% computation of the system stiffness matrix and force vector
[stiffness]=...
    formStiffnessMatrixMindlinQ4(GDof,numberElements,...
    elementNodes,numberNodes,nodeCoordinates,C_shear,...
    C_bending,thickness,I);

[force]=...
    formForceVectorMindlinQ4(GDof,numberElements,...
    elementNodes,numberNodes,nodeCoordinates,P);

% % boundary conditions
[prescribedDof,activeDof]=...
    EssentialBC('ssss',GDof,xx,yy,nodeCoordinates,numberNodes);

% solution
displacements=solution(GDof,prescribedDof,stiffness,force);

% displacements
disp('Displacements')
jj=1:GDof; format
f=[jj; displacements'];
fprintf('node U\n')
fprintf('%3d %12.8f\n',f)

% deformed shape
figure
plot3(xx,yy,displacements(1:numberNodes),'.')
format long
D1=E*thickness^3/12/(1-poisson^2);
min(displacements(1:numberNodes))*D1/L^4
```

This MATLAB code calls functions formStiffnessMatrixMindlinQ4.m for computation of stiffness matrix and formForceVectorMindlinQ4.m for computation of the force vector.

```matlab
%.....................................................................

function [K]=...
    formStiffnessMatrixMindlinQ4(GDof,numberElements,...
    elementNodes,numberNodes,nodeCoordinates,C_shear,...
    C_bending,thickness,I)

% computation of stiffness matrix
% for Mindlin plate element

% K : stiffness matrix

K=zeros(GDof);

% Gauss quadrature for bending part
[gaussWeights,gaussLocations]=gaussQuadrature('complete');

% cycle for element
% cycle for element
for e=1:numberElements
  % indice : nodal condofectivities for each element
  % elementDof: element degrees of freedom
  indice=elementNodes(e,:);
  elementDof=[indice indice+numberNodes indice+2*numberNodes];
  ndof=length(indice);

  % cycle for Gauss point
  for q=1:size(gaussWeights,1)
    GaussPoint=gaussLocations(q,:);
    xi=GaussPoint(1);
    eta=GaussPoint(2);

% shape functions and derivatives
    [shapeFunction,naturalDerivatives]=shapeFunctionQ4(xi,eta)

% Jacobian matrix, inverse of Jacobian,
% derivatives w.r.t. x,y
    [Jacob,invJacobian,XYderivatives]=...
        Jacobian(nodeCoordinates(indice,:),naturalDerivatives);
```

```
% [B] matrix bending
    B_b=zeros(3,3*ndof);
    B_b(1,ndof+1:2*ndof)  = XYderivatives(:,1)';
    B_b(2,2*ndof+1:3*ndof)= XYderivatives(:,2)';
    B_b(3,ndof+1:2*ndof)  = XYderivatives(:,2)';
    B_b(3,2*ndof+1:3*ndof)= XYderivatives(:,1)';

% stiffness matrix bending
    K(elementDof,elementDof)=K(elementDof,elementDof)+ ...
        B_b'*C_bending*B_b*gaussWeights(q)*det(Jacob);
    end  % Gauss point
end     % element

% shear stiffness matrix

% Gauss quadrature for shear part
[gaussWeights,gaussLocations]=gaussQuadrature('reduced');

% cycle for element
% cycle for element
for e=1:numberElements
  % indice : nodal condofectivities for each element
  % elementDof: element degrees of freedom
  indice=elementNodes(e,:);
  elementDof=[indice indice+numberNodes indice+2*numberNodes];
  ndof=length(indice);

  % cycle for Gauss point
  for q=1:size(gaussWeights,1)
    GaussPoint=gaussLocations(q,:);
    xi=GaussPoint(1);
    eta=GaussPoint(2);

% shape functions and derivatives
    [shapeFunction,naturalDerivatives]=shapeFunctionQ4(xi,eta)

% Jacobian matrix, inverse of Jacobian,
% derivatives w.r.t. x,y
    [Jacob,invJacobian,XYderivatives]=...
        Jacobian(nodeCoordinates(indice,:),naturalDerivatives);
% [B] matrix shear
    B_s=zeros(2,3*ndof);
```

```
    B_s(1,1:ndof)        = XYderivatives(:,1)';
    B_s(2,1:ndof)        = XYderivatives(:,2)';
    B_s(1,ndof+1:2*ndof) = shapeFunction;
    B_s(2,2*ndof+1:3*ndof)= shapeFunction;

% stiffness matrix shear
    K(elementDof,elementDof)=K(elementDof,elementDof)+...
        B_s'*C_shear  *B_s*gaussWeights(q)*det(Jacob);
  end  % gauss point
end    % element
```

```
%..................................................................

function [force]=...
    formForceVectorMindlinQ4(GDof,numberElements,...
    elementNodes,numberNodes,nodeCoordinates,P)

% computation of force vector
% for Mindlin plate element

% force : force vector
force=zeros(GDof,1);

% Gauss quadrature for bending part
[gaussWeights,gaussLocations]=gaussQuadrature('reduced');

% cycle for element
for e=1:numberElements
  % indice : nodal connectivities for each element
  indice=elementNodes(e,:);

  % cycle for Gauss point
  for q=1:size(gaussWeights,1)
    GaussPoint=gaussLocations(q,:);
    GaussWeight=gaussWeights(q);
    xi=GaussPoint(1);
    eta=GaussPoint(2);

% shape functions and derivatives
    [shapeFunction,naturalDerivatives]=shapeFunctionQ4(xi,eta)
```

```
% Jacobian matrix, inverse of Jacobian,
% derivatives w.r.t. x,y
    [Jacob,invJacobian,XYderivatives]=...
        Jacobian(nodeCoordinates(indice,:),naturalDerivatives);

% force vector
force(indice)=force(indice)+shapeFunction*P*det(Jacob)
    *GaussWeight;
  end  % Gauss point

end     % element
```

The imposition of the essential boundary conditions is made in function Essen-tialBC.m, as:

```
%................................................................

function [prescribedDof,activeDof,fixedNodeW]=...
    EssentialBC(typeBC,GDof,xx,yy,nodeCoordinates,numberNodes)

% essentialBoundary conditions for recatngular plates
switch typeBC
    case 'ssss'
fixedNodeW =find(xx==max(nodeCoordinates(:,1))|...
                 xx==min(nodeCoordinates(:,1))|...
                 yy==min(nodeCoordinates(:,2))|...
                 yy==max(nodeCoordinates(:,2)));

fixedNodeTX =find(yy==max(nodeCoordinates(:,2))|...
    yy==min(nodeCoordinates(:,2)));
fixedNodeTY =find(xx==max(nodeCoordinates(:,1))|...
    xx==min(nodeCoordinates(:,1)));
    case 'cccc'

fixedNodeW =find(xx==max(nodeCoordinates(:,1))|...
                 xx==min(nodeCoordinates(:,1))|...
                 yy==min(nodeCoordinates(:,2))|...
                 yy==max(nodeCoordinates(:,2)));
fixedNodeTX =fixedNodeW;
fixedNodeTY =fixedNodeTX;
    case 'scsc'
```

```
fixedNodeW =find(xx==max(nodeCoordinates(:,1))|...
               xx==min(nodeCoordinates(:,1))|...
               yy==min(nodeCoordinates(:,2))|...
               yy==max(nodeCoordinates(:,2)));

fixedNodeTX =find(xx==max(nodeCoordinates(:,2))|...
    xx==min(nodeCoordinates(:,2)));
fixedNodeTY=[];
    case 'cccf'
fixedNodeW =find(xx==min(nodeCoordinates(:,1))|...
               yy==min(nodeCoordinates(:,2))|...
               yy==max(nodeCoordinates(:,2)));

fixedNodeTX =fixedNodeW;
fixedNodeTY =fixedNodeTX;
end

prescribedDof=[fixedNodeW;fixedNodeTX+numberNodes;...
    fixedNodeTY+2*numberNodes]
activeDof=setdiff([1:GDof]',[prescribedDof]);
```

The Mindlin plate problem19 can be solved using MATLAB structures. Code problem19structure.m includes all the functions needed for the execution of the problem.

```
%.............................................................
% MATLAB codes for Finite Element Analysis
% problem19structure.m
% Mindlin plate in bending
% antonio ferreira 2008
function problem19structure
% clear memory
clear all;colordef white;clf

% materials
E  = 10920;      poisson = 0.30; kapa=5/6;
thickness=0.001;
I=thickness^3/12;

% load
```

```
element=struct('P',-1);

%Mesh generation
L = 1;
numberElementsX=20; numberElementsY=20;
element.numberElements=numberElementsX*numberElementsY;
[element.nodeCoordinates, element.elementNodes] = ...
    rectangularMesh(L,L,numberElementsX,numberElementsY);
element.numberNodes=size(element.nodeCoordinates,1);
element.GDof=3*size(element.nodeCoordinates,1);

% matrix C
% bending part :C_bending
% shear part : C_shear
element.C_bending=...
    I*E/(1-poisson^2)*[1 poisson 0;poisson 1 0;0 0 (1-poisson)/2];
element.C_shear=...
    kapa*thickness*E/2/(1+poisson)*eye(2);

% computation of the system stiffness matrix and force vector
element.stiffness=formStiffnessMatrixMindlinQ4Structure(element);
element.force=formForceVectorMindlinQ4Structure(element);

% % boundary conditions
element.prescribed=EssentialBCStructure('ssss',element);

% solution
element.displacements=solutionStructure(element);

% displacements
disp('Displacements')
jj=1:element.GDof;
f=[jj; element.displacements'];
fprintf('node U\n')
fprintf('%3d %12.8f\n',f)

% original mesh
drawingMesh(element.nodeCoordinates,element.elementNodes,'Q4',
    'k-');
axis off

% deformed shape
figure
```

```
plot3(element.nodeCoordinates(:,1),element.nodeCoordinates(:,2)...
    ,element.displacements(1:size(element.nodeCoordinates,1)),'.')
format long
D1=E*thickness^3/12/(1-poisson^2);
min(element.displacements(1:size(element.nodeCoordinates,1)))
    *D1/L^4


end

%.............................................................
function [K]=formStiffnessMatrixMindlinQ4Structure(element)

% computation of stiffness matrix
% for Mindlin plate element

% K : stiffness matrix
K=zeros(element.GDof);

% Gauss quadrature for bending part
[quadrature]=getQuadratureStructure;

% cycle for element
for e=1:element.numberElements
  % indice : nodal conectivities for each element
  indice=element.elementNodes(e,:);
  % indice : nodal conectivities for each element
  % elementDof: element degrees of freedom
  elementDof=[indice indice+element.numberNodes...
      indice+2*element.numberNodes];
  ndof=length(indice);

  % cycle for Gauss point
  for q=1:size(quadrature(2).weights,1)
    GaussPoint=quadrature(2).points
    xi=GaussPoint(1);
    eta=GaussPoint(2);

% shape functions and derivatives
    [shapeFunction]=getShapeFunctionStructure(xi,eta)

% Jacobian matrix, inverse of Jacobian,
% derivatives w.r.t. x,y
```

```
[Jac]=JacobianStructure(element.nodeCoordinates(indice,:),...
    shapeFunction(1).naturalDerivatives)

% [B] matrix bending
    B_b=zeros(3,3*ndof);
    B_b(1,ndof+1:2*ndof)  = Jac.derivatives(:,1)';
    B_b(2,2*ndof+1:3*ndof)= Jac.derivatives(:,2)';
    B_b(3,ndof+1:2*ndof)  = Jac.derivatives(:,2)';
    B_b(3,2*ndof+1:3*ndof)= Jac.derivatives(:,1)';

% stiffness matrix bending
    K(elementDof,elementDof)=K(elementDof,elementDof)+ ...
        B_b'*element.C_bending*B_b*quadrature(2).weights(q)*...
        det(Jac.matrix);
    end % Gauss point
end    % element

% shear stiffness matrix

% cycle for element
for e=1:element.numberElements
  % indice : nodal conectivities for each element
  indice=element.elementNodes(e,:)    ;
  % indice : nodal conectivities for each element
  % elementDof: element degrees of freedom
  elementDof=[indice indice+element.numberNodes...
      indice+2*element.numberNodes];
  ndof=length(indice);

  % cycle for Gauss point ! one Gauss point (reduced)
  for q=1:size(quadrature(1).weights,1)
    GaussPoint=quadrature(1).points
    xi=GaussPoint(1);
    eta=GaussPoint(2);

% shape functions and derivatives
    [shapeFunction]=getShapeFunctionStructure(xi,eta)

% Jacobian matrix, inverse of Jacobian,
% derivatives w.r.t. x,y

[Jac]=JacobianStructure(element.nodeCoordinates(indice,:),...
    shapeFunction(1).naturalDerivatives)
```

```
% [B] matrix shear
    B_s=zeros(2,3*ndof);
    B_s(1,1:ndof)        = Jac.derivatives(:,1)';
    B_s(2,1:ndof)        = Jac.derivatives(:,2)';
    B_s(1,ndof+1:2*ndof) = shapeFunction(1).shape;
    B_s(2,2*ndof+1:3*ndof)= shapeFunction(1).shape;

% stiffness matrix shear
    K(elementDof,elementDof)=K(elementDof,elementDof)+...
        B_s'*element.C_shear*B_s*quadrature(1).weights(q)*...
        det(Jac.matrix);
  end  % gauss point
end    % element


end

%....................................................
function [force]=formForceVectorMindlinQ4Structure(element)

% computation of force vector
% for Mindlin plate element

% force : force vector
force=zeros(element.GDof,1);

% Gauss quadrature for bending part
[quadrature]=getQuadratureStructure;

% cycle for element
for e=1:element.numberElements
  % indice : nodal conectivities for each element
  indice=element.elementNodes(e,:)    ;

  % cycle for Gauss point
  for q=1:size(quadrature(2).weights,1)
    GaussPoint=quadrature(2).points
    xi=GaussPoint(1);
    eta=GaussPoint(2);

% shape functions and derivatives
    [shapeFunction]=getShapeFunctionStructure(xi,eta)
```

```
% Jacobian matrix, inverse of Jacobian,
% derivatives w.r.t. x,y

    [Jac]=JacobianStructure(element.nodeCoordinates(indice,:),...
    shapeFunction(1).naturalDerivatives)


% force vector
    force(indice)=force(indice)+...
        shapeFunction(1).shape(q)*element.P...
        *det(Jac.matrix)*quadrature(2).weights(q);
  end  % Gauss point
end    % element
end

% .........................................................
    function [quadrature]=getQuadratureStructure

    % quadrature points and weights for Gauss quadrature
    % quadrilaterals and triangles

    % points: Gauss points
    % weights: Gauss weights

    % Structure quadrature
    quadrature=struct()

    % order = 1 (1 x 1)
    quadrature(1).points=[0;0];
    quadrature(1).weights=4;

    % order = 2 (2 x 2)
    quadrature(2).points=[...
            -0.577350269189626 -0.577350269189626;
             0.577350269189626 -0.577350269189626;
             0.577350269189626  0.577350269189626;
            -0.577350269189626  0.577350269189626];
    quadrature(2).weights=[ 1;1;1;1];

    % order = 3 (3 x 3)
    quadrature(3).points=[...
            -0.774596669241483 -0.774596669241483;
            -0.774596669241483  0.0;
```

```
                -0.774596669241483  0.774596669241483;
                0. -0.774596669241483;
                0.  0.0;
                0.  0.774596669241483;
                0.774596669241483 -0.774596669241483;
                0.774596669241483  0.0;
                0.774596669241483  0.774596669241483];
  quadrature(3).weights=[0.555555555555556*0.555555555555556;
                         0.555555555555556*0.888888888888889;
                         0.555555555555556*0.555555555555556;
                         0.55555555555556*0.888888888888889;
                         0.888888888888889*0.888888888888889;
                         0.555555555555556*0.555555555555556;
                         0.555555555555556*0.555555555555556;
                         0.555555555555556*0.888888888888889;
                         0.555555555555556*0.555555555555556];


  % order = 4 Triangles (1 point)
  quadrature(4).points=[ 0.3333333333333, 0.3333333333333 ];
  quadrature(4).weights=[1/2];

  % order = 5 Triangles (3 points)
  quadrature(5).points= [ 0.1666666666667, 0.1666666666667 ;
                          0.6666666666667, 0.1666666666667 ;
                          0.1666666666667, 0.6666666666667 ];
  quadrature(5).weights=[1/3;1/3;1/3];
end % end function getQuadrature

% ............................................................
function [shapeFunction]=getShapeFunctionStructure(xi,eta)

% shape function and derivatives for Q4,T3,Q9 and Q8 elements
% shape : Shape functions
% naturalDerivatives: derivatives w.r.t. xi and eta
% xi, eta: natural coordinates (-1 ... +1)

% Structure shapeFunction
shapeFunction=struct()

% Q4 element
shapeFunction(1).shape=1/4*[ ...
                            (1-xi)*(1-eta);(1+xi)*(1-eta);
                            (1+xi)*(1+eta);(1-xi)*(1+eta)];
```

```
        shapeFunction(1).naturalDerivatives=...
                    1/4*[-(1-eta), -(1-xi);1-eta,    -(1+xi);
                          1+eta,      1+xi;-(1+eta),   1-xi];
% T3 element
shapeFunction(2).shape=[1-xi-eta;xi;eta];
shapeFunction(2).naturalDerivatives=[-1,-1;1,0;0,1];

% Q9 element
shapeFunction(3).shape=1/4*[xi*eta*(xi-1)*(eta-1);
          xi*eta*(xi+1)*(eta-1);
          xi*eta*(xi+1)*(eta+1);
          xi*eta*(xi-1)*(eta+1);
         -2*eta*(xi+1)*(xi-1)*(eta-1);
         -2*xi*(xi+1)*(eta+1)*(eta-1);
         -2*eta*(xi+1)*(xi-1)*(eta+1);
         -2*xi*(xi-1)*(eta+1)*(eta-1);
          4*(xi+1)*(xi-1)*(eta+1)*(eta-1)];
shapeFunction(3).naturalDerivatives=...
    1/4*[eta*(2*xi-1)*(eta-1),xi*(xi-1)*(2*eta-1);
            eta*(2*xi+1)*(eta-1),xi*(xi+1)*(2*eta-1);
            eta*(2*xi+1)*(eta+1),xi*(xi+1)*(2*eta+1);
            eta*(2*xi-1)*(eta+1),xi*(xi-1)*(2*eta+1);
          -4*xi*eta*(eta-1),   -2*(xi+1)*(xi-1)*(2*eta-1);
   -2*(2*xi+1)*(eta+1)*(eta-1),-4*xi*eta*(xi+1);
          -4*xi*eta*(eta+1),   -2*(xi+1)*(xi-1)*(2*eta+1);
   -2*(2*xi-1)*(eta+1)*(eta-1),-4*xi*eta*(xi-1);
          8*xi*(eta^2-1),       8*eta*(xi^2-1)];

% Q8 element
shapeFunction(4).shape=[1/4*xi*(1-xi)*eta*(1-eta);
   -1/2*xi*(1-xi)*(1+eta)*(1-eta);
   -1/4*xi*(1-xi)*eta*(1+eta);
   1/2*(1+xi)*(1-xi)*(1+eta)*eta;
   1/4*xi*(1+xi)*eta*(1+eta);
   1/2*xi*(1+xi)*(1+eta)*(1-eta);
   -1/4*xi*(1+xi)*eta*(1-eta);
   -1/2*(1+xi)*(1-xi)*(1-eta)*eta];
shapeFunction(4).naturalDerivatives=...
  [1/4*eta*(-1+eta)*(-1+2*xi)  1/4*xi*(-1+xi)*(-1+2*eta);
    -1/2*(1+eta)*(-1+eta)*(-1+2*xi)   -xi*(-1+xi)*eta;
    1/4*eta*(1+eta)*(-1+2*xi)   1/4*xi*(-1+xi)*(1+2*eta);
    -xi*eta*(1+eta) -1/2*(1+xi)*(-1+xi)*(1+2*eta);
    1/4*eta*(1+eta)*(1+2*xi) 1/4*xi*(1+xi)*(1+2*eta);
```

```
            -1/2*(1+eta)*(-1+eta)*(1+2*xi) -xi*(1+xi)*eta;
            1/4*eta*(-1+eta)*(1+2*xi) 1/4*xi*(1+xi)*(-1+2*eta);
            -xi*eta*(-1+eta) -1/2*(1+xi)*(-1+xi)*(-1+2*eta)];

    end % end function shapeFunctionQ4

    % ..........................................................
function [Jac]=JacobianStructure(nodeCoordinates,natural
    Derivatives)

    % Jac.matrix     : Jacobian matrix
    % Jac.inv : inverse of Jacobian Matrix
    % Jac.derivatives  : derivatives w.r.t. x and y
    % naturalDerivatives  : derivatives w.r.t. xi and eta
    % nodeCoordinates  : nodal coordinates at element level

    Jac=struct();

    Jac.matrix=nodeCoordinates'*naturalDerivatives;
    Jac.inv=inv(Jac.matrix);
    Jac.derivatives=naturalDerivatives*Jac.inv;

    end % end function Jacobian


    %..........................................................

function prescribed=EssentialBCStructure(typeBC,element)

% essentialBoundary conditions for rectangular plates
xx=element.nodeCoordinates(:,1);
yy=element.nodeCoordinates(:,2);

switch typeBC
    case 'ssss'
fixedNodeW =find(xx==max(element.nodeCoordinates(:,1))|...
                xx==min(element.nodeCoordinates(:,1))|...
                yy==min(element.nodeCoordinates(:,2))|...
                yy==max(element.nodeCoordinates(:,2)));

fixedNodeTX =find(yy==max(element.nodeCoordinates(:,2))|...
    yy==min(element.nodeCoordinates(:,2)));
fixedNodeTY =find(xx==max(element.nodeCoordinates(:,1))|...
```

```
        xx==min(element.nodeCoordinates(:,1)));
        case 'cccc'

    fixedNodeW =find(xx==max(element.nodeCoordinates(:,1))|...
                    xx==min(element.nodeCoordinates(:,1))|...
                    yy==min(element.nodeCoordinates(:,2))|...
                    yy==max(element.nodeCoordinates(:,2)));
    fixedNodeTX =fixedNodeW;
    fixedNodeTY =fixedNodeTX;
        case 'scsc'

    fixedNodeW =find(xx==max(element.nodeCoordinates(:,1))|...
                    xx==min(element.nodeCoordinates(:,1))|...
                    yy==min(element.nodeCoordinates(:,2))|...
                    yy==max(element.nodeCoordinates(:,2)));

    fixedNodeTX =find(xx==max(element.nodeCoordinates(:,2))|...
        xx==min(element.nodeCoordinates(:,2)));
    fixedNodeTY=[];
        case 'cccf'
    fixedNodeW =find(xx==min(element.nodeCoordinates(:,1))|...
                    yy==min(element.nodeCoordinates(:,2))|...
                    yy==max(element.nodeCoordinates(:,2)));

    fixedNodeTX =fixedNodeW;
    fixedNodeTY =fixedNodeTX;
    end

    prescribed=[fixedNodeW;fixedNodeTX+element.numberNodes;...
        fixedNodeTY+2*element.numberNodes];
    end


%............................................................

function displacements=solutionStructure(element)
% function to find solution in terms of global displacements

activeDof=setdiff([1:element.GDof]', ...
    [element.prescribed]);
U=element.stiffness(activeDof,activeDof)\...
    element.force(activeDof);
displacements=zeros(element.GDof,1);
```

```
displacements(activeDof)=U;
end
```

12.5 Free vibrations of Mindlin plates

By using the Hamilton principle [15], we may express the equations of motion of Mindlin plates as

$$\mathbf{M}\ddot{\mathbf{u}} + \mathbf{K}\mathbf{u} = \mathbf{f} \qquad (12.28)$$

where $\mathbf{M}, \mathbf{K}, \mathbf{f}$ are the system mass and stiffness matrices, and the force vector, respectively, and $\ddot{\mathbf{u}}, \mathbf{u}$ are the accelerations and displacements. Assuming a harmonic motion we obtain the natural frequencies and the modes of vibration by solving the generalized eigenproblem [8]

$$\left(\mathbf{K} - \omega^2 \mathbf{M}\right)\mathbf{X} = \mathbf{0} \qquad (12.29)$$

where ω is the natural frequency and $\mathbf{X}$ the mode of vibration.

By using the kinetic energy for the plate

$$T^e = \frac{1}{2}\int_A \rho\left[h\dot{w}^2 + \frac{h^3}{12}\dot{\theta}_x^2 + \frac{h^3}{12}\dot{\theta}_y^2\right]dA \qquad (12.30)$$

we may compute the mass matrix as [15]

$$\mathbf{M}^e = \int_A \rho\mathbf{N}^T \begin{bmatrix} h & 0 & 0 \\ 0 & \frac{h^3}{12} & 0 \\ 0 & 0 & \frac{h^3}{12} \end{bmatrix}\mathbf{N}dA \qquad (12.31)$$

being the stiffness matrix already obtained before for static problems.

We consider a square plate (side a), with thickness-to-side ratio $h/a = 0.01$ and $h/a = 0.1$. The non-dimensional natural frequency is given by

$$\bar{\omega} = \omega_{mn}a\sqrt{\frac{\rho}{G}},$$

where ρ is the material density, G the shear modulus ($G = E/(2(1 + \nu))$), E the modulus of elasticity and ν the Poisson's coefficient. Indices m and n are the vibration half-waves in axes x and y. In this problem we consider fully simply-supported (SSSS) and fully clamped (CCCC) plates, as well as SCSC and CCCF plates where F means free side.

For CCCC and CCCF we use a shear correction factor $k = 0.8601$, while for SCSC plates we use $k = 0.822$. For SSSS plates we consider $k = 5/6$.

In table 12.2 we show the convergence of the fundamental frequency for CCCC plate with $h/a = 0.01, k = 0.8601, \nu = 0.3$. We obtain quite good agreement with the analytical solution [19].

In table 12.3 we show the convergence of the fundamental frequency for SSSS plate with $h/a = 0.01, k = 0.8333, \nu = 0.3$. Again, we obtain quite good agreement with a analytical solution [19]. Tables 12.4 and 12.5 consider $h/a = 1$ and in all of them results agree very well with analytical solution.

Tables 12.6 and 12.7 list the natural frequencies of a SSSS plate with $h/a = 0.1$ and $h/a = 0.01$, being $k = 0.833, \nu = 0.3$. Our finite element solution agrees with the tridimensional solution and analytical solution given by Mindlin [8].

Tables 12.8 and 12.9 compare natural frequencies with the Rayleygh-Ritz solution [8] and a solution by Liew [20].

Tables 12.10 and 12.11 compare natural frequencies for SCSC plate with $h/a = 0.1$ and $h/a = 0.01$, being $k = 0.822, \nu = 0.3$, respectively. Sides located at $x = 0; L$ are simply-supported.

Table 12.2 Convergence of natural frequency $\bar{\omega}$ for CCCC plate with $h/a = 0.01, k = 0.8601, \nu = 0.3$

10 × 10 Q4	0.1800	Analytical [19] 0.1754
15 × 15 Q4	0.1774	
20 × 20 Q4	0.1765	
25 × 25 Q4	0.1761	

Table 12.3 Convergence of natural frequency $\bar{\omega}$ for SSSS plate with $h/a = 0.01, k = 0.8333, \nu = 0.3$

10 × 10 Q4	0.0973	Analytical [19] 0.0963
15 × 15 Q4	0.0968	
20 × 20 Q4	0.0965	
25 × 25 Q4	0.0965	

Table 12.4 Convergence of natural frequency $\bar{\omega}$ for CCCC plate with $h/a = 0.1, k = 0.8601, \nu = 0.3$

10 × 10 Q4	1.6259	Analytical [19] 1.5940
15 × 15 Q4	1.6063	
20 × 20 Q4	1.5996	

Table 12.5 Convergence of natural frequency $\bar{\omega}$ for SSSS plate with $h/a = 0.1, k = 0.8333, \nu = 0.3$

10 × 10 Q4	0.9399	Analytical [19] 0.930
15 × 15 Q4	0.9346	
20 × 20 Q4	0.9327	

Table 12.6 Natural frequencies of a SSSS plate with $h/a = 0.1, k = 0.833, \nu = 0.3$

Mode no.	m	n	15 × 15 Q4	3D *	Mindlin *
1	1	1	0.9346	0.932	0.930
2	2	1	2.2545	2.226	2.219
3	1	2	2.2545	2.226	2.219
4	2	2	3.4592	3.421	3.406
5	3	1	4.3031	4.171	4.149
6	1	3	4.3031	4.171	4.149
7	3	2	5.3535	5.239	5.206
8	2	3	5.3535	5.239	5.206
9	4	1	6.9413	–	6.520
10	1	4	6.9413	–	6.520
11	3	3	7.0318	6.889	6.834
12	4	2	7.8261	7.511	7.446
13	2	4	7.8261	7.511	7.446

* analytical solution

Table 12.7 Natural frequencies of a SSSS plate with $h/a = 0.01, k = 0.833, \nu = 0.3$

Mode no.	m	n	20 × 20 Q4	Mindlin *
1	1	1	0.0965	0.0963
2	2	1	0.2430	0.2406
3	1	2	0.2430	0.2406
4	2	2	0.3890	0.3847
5	3	1	0.4928	0.4807
6	1	3	0.4928	0.4807
7	3	2	0.6380	0.6246
8	2	3	0.6380	0.6246
9	4	1	0.8550	0.8156
10	1	4	0.8550	0.8156
11	3	3	0.8857	0.8640
12	4	2	0.9991	0.9592
13	2	4	0.9991	0.9592

(* – analytical solution)

Tables 12.12 and 12.13 compare natural frequencies for CCCF plates with $h/a = 0.1$ and $h/a = 0.01$, respectively, being $k = 0.822, \nu = 0.3$. Side located at $x = L$ is free.

The present finite element results are quite accurate.

Figure 12.3 shows the modes of vibration for a CCCC plate with $h/a = 0.1$, using 10 × 10 Q4 elements.

Figure 12.4 shows the modes of vibration for a SSSS plate with $h/a = 0.1$, using 10 × 10 Q4 elements.

Figure 12.5 shows the modes of vibration for a SCSC plate with $h/a = 0.01$, using 15 × 15 Q4 elements.

Table 12.8 Natural frequencies of a CCCC plate with $h/a = 0.1, k = 0.8601, \nu = 0.3$

Mode no.	m	n	20 × 20 Q4	Rayleygh-Ritz [19]	Liew et al. [20]
1	1	1	1.5955	1.5940	1.5582
2	2	1	3.0662	3.0390	3.0182
3	1	2	3.0662	3.0390	3.0182
4	2	2	4.2924	4.2650	4.1711
5	3	1	5.1232	5.0350	5.1218
6	1	3	5.1730	5.0780	5.1594
7	3	2	6.1587		6.0178
8	2	3	6.1587		6.0178
9	4	1	7.6554		7.5169
10	1	4	7.6554		7.5169
11	3	3	7.7703		7.7288
12	4	2	8.4555		8.3985
13	2	4	8.5378		8.3985

Table 12.9 Natural frequencies of a CCCC plate with $h/a = 0.01, k = 0.8601, \nu = 0.3$

Mode no.	m	n	20 × 20 Q4	Rayleygh-Ritz [19]	Liew et al. [20]
1	1	1	0.175	0.1754	0.1743
2	2	1	0.3635	0.3576	0.3576
3	1	2	0.3635	0.3576	0.3576
4	2	2	0.5358	0.5274	0.5240
5	3	1	0.6634	0.6402	0.6465
6	1	3	0.6665	0.6432	0.6505
7	3	2	0.8266		0.8015
8	2	3	0.8266		0.8015
9	4	1	1.0875		1.0426
10	1	4	1.0875		1.0426
11	3	3	1.1049		1.0628
12	4	2	1.2392		1.1823
13	2	4	1.2446		1.1823

Table 12.10 Natural frequencies for SCSC plate with $h/a = 0.1, k = 0.822, \nu = 0.3$

Mode no.	m	n	15 × 15 Q4	Mindlin solution [8]
1	1	1	1.2940	1.302
2	2	1	2.3971	2.398
3	1	2	2.9290	2.888
4	2	2	3.8394	3.852
5	3	1	4.3475	4.237
6	1	3	5.1354	4.936
7	3	2	5.5094	
8	2	3	5.8974	
9	4	1	6.9384	
10	1	4	7.2939	
11	3	3	7.7968	
12	4	2	7.8516	
13	2	4	8.4308	

Table 12.11 Natural frequencies for SCSC plate with $h/a = 0.01, k = 0.822, \nu = 0.3$

Mode no.	m	n	15 × 15 Q4	Mindlin solution [8]
1	1	1	0.1424	0.1411
2	2	1	0.2710	0.2668
3	1	2	0.3484	0.3377
4	2	2	0.4722	0.4608
5	3	1	0.5191	0.4979
6	1	3	0.6710	0.6279
7	3	2	0.7080	
8	2	3	0.7944	
9	4	1	0.8988	
10	1	4	1.0228	
11	3	3	1.0758	
12	4	2	1.1339	
13	2	4	1.2570	

Table 12.12 Natural frequencies for CCCF plate with $h/a = 0.1, k = 0.8601, \nu = 0.3$

Mode no.	m	n	15 × 15 Q4	Mindlin solution [8]
1	1	1	1.0923	1.089
2	2	1	1.7566	1.758
3	1	2	2.7337	2.673
4	2	2	3.2591	3.216
5	3	1	3.3541	3.318
6	1	3	4.6395	4.615
7	3	2	4.9746	
8	2	3	5.4620	
9	4	1	5.5245	
10	1	4	6.5865	
11	3	3	6.6347	
12	4	2	7.6904	
13	2	4	8.1626	

Table 12.13 Natural frequencies for CCCF plate with $h/a = 0.01, k = 0.8601, \nu = 0.3$

Mode no.	m	n	15 × 15 Q4	Mindlin solution [8]
1	1	1	0.1180	0.1171
2	2	1	0.1967	0.1951
3	1	2	0.3193	0.3093
4	2	2	0.3830	0.3740
5	3	1	0.4031	0.3931
6	1	3	0.5839	0.5695
7	3	2	0.6387	
8	2	3	0.7243	
9	4	1	0.8817	
10	1	4	0.9046	
11	3	3	1.0994	
12	4	2	1.1407	
13	2	4	1.1853	

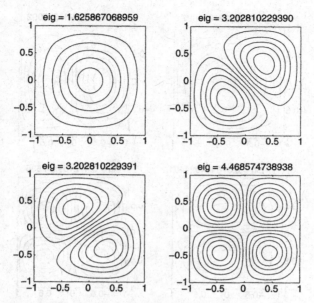

Fig. 12.3 Modes of vibration for a CCCC plate with $h/a = 0.1$, using 10×10 Q4 elements

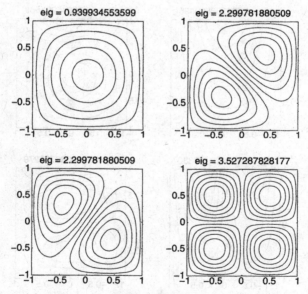

Fig. 12.4 Modes of vibration for a SSSS plate with $h/a = 0.1$, using 10×10 Q4 elements

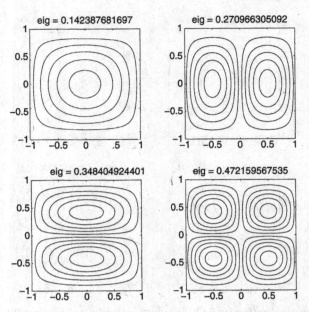

Fig. 12.5 Modes of vibration for a SCSC plate with $h/a = 0.01$, using 15×15 Q4 elements

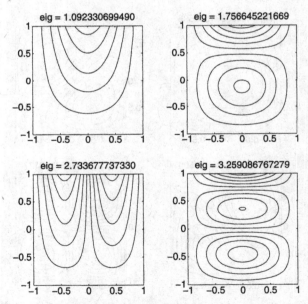

Fig. 12.6 Modes of vibration for a CCCF plate with $h/a = 0.1$, using 15×15 Q4 elements

Figure 12.6 shows the modes of vibration for a CCCF plate with $h/a = 0.1$, using 15×15 Q4 elements.

The MATLAB code (problem19Vibrations.m), solves the free vibration problem of Mindlin plates. The user is requested to change input details according to the problem.

```
%..............................................................

% MATLAB codes for Finite Element Analysis
% problem19vibrations.m
% Mindlin plate in free vibrations
% antonio ferreira 2008

% clear memory
clear all;colordef white;clf

% materials
E  = 10920;       poisson = 0.30;
thickness=0.1;
I=thickness^3/12;
rho=1;
kapa=0.8601; % cccc / cccf case
%kapa=0.822; % scsc case
kapa=5/6;  % ssss case

% matrix C
% bending part
C_bending=I*E/(1-poisson^2)*...
    [1 poisson 0;poisson 1 0;0 0 (1-poisson)/2];
% shear part
C_shear=kapa*thickness*E/2/(1+poisson)*eye(2);

% load
P = -1;

%Mesh generation
L  = 1;
numberElementsX=10;
numberElementsY=10;
numberElements=numberElementsX*numberElementsY;

[nodeCoordinates, elementNodes] = ...
    rectangularMesh(L,L,numberElementsX,numberElementsY);
xx=nodeCoordinates(:,1);
yy=nodeCoordinates(:,2);
```

```
drawingMesh(nodeCoordinates,elementNodes,'Q4','k-');
axis off
numberNodes=size(xx,1);

% GDof: global number of degrees of freedom
GDof=3*numberNodes;

% computation of the system stiffness and mass matrices
[stiffness]=...
    formStiffnessMatrixMindlinQ4(GDof,numberElements,...
    elementNodes,numberNodes,nodeCoordinates,C_shear,...
    C_bending,thickness,I);

[mass]=...
formMassMatrixMindlinQ4(GDof,numberElements,...
elementNodes,numberNodes,nodeCoordinates,thickness,rho,I);

% % boundary conditions
[prescribedDof,activeDof,fixedNodeW]=...
EssentialBC('cccc',GDof,xx,yy,nodeCoordinates,numberNodes)

G=E/2.6;
% V : mode shape
% D : frequency
%
numberOfModes=12;
[V,D] = eig(stiffness(activeDof,activeDof),...
    mass(activeDof,activeDof));
D = diag(sqrt(D)*L*sqrt(rho/G));
[D,ii] = sort(D); ii = ii(1:numberOfModes);
VV = V(:,ii);
activeDofW=setdiff([1:numberNodes]',[fixedNodeW]);
NNN=size(activeDofW);

    VVV(1:numberNodes,1:12)=0;
    for i=1:12
        VVV(activeDofW,i)=VV(1:NNN,i);
    end

NN=numberNodes;N=sqrt(NN);
x=linspace(-L,L,numberElementsX+1);
y=linspace(-L,L,numberElementsY+1);
```

```
% drawing Eigenmodes
drawEigenmodes2D(x,y,VVV,NN,N,D)
```

This code calls function formMassMatrixMindlinQ4.m which computes the mass matrices of the Mindlin Q4 element. The code for computing the stiffness matrix has already been presented.

```
%..............................................................
function [mass]=...
    formMassMatrixMindlinQ4(GDof,numberElements,...
    elementNodes,numberNodes,nodeCoordinates,thickness,rho,I)

% computation of  mass matrix
% for Mindlin plate element

% mass : mass matrix
mass=zeros(GDof);

% Gauss quadrature for bending part
[gaussWeights,gaussLocations]=gaussQuadrature('complete');

% cycle for element
for e=1:numberElements
  % indice : nodal condofectivities for each element
  indice=elementNodes(e,:);
  ndof=length(indice);

  % cycle for Gauss point
  for q=1:size(gaussWeights,1)
    GaussPoint=gaussLocations(q,:);
    xi=GaussPoint(1);
    eta=GaussPoint(2);

% shape functions and derivatives
    [shapeFunction,naturalDerivatives]=shapeFunctionQ4(xi,eta)

% Jacobian matrix, inverse of Jacobian,
```

```
% derivatives w.r.t. x,y
    [Jacob,invJacobian,XYderivatives]=...
        Jacobian(nodeCoordinates(indice,:),naturalDerivatives);

% [B] matrix bending
    B_b=zeros(3,3*ndof);
    B_b(1,ndof+1:2*ndof)   = XYderivatives(:,1)';
    B_b(2,2*ndof+1:3*ndof)= XYderivatives(:,2)';
    B_b(3,ndof+1:2*ndof)   = XYderivatives(:,2)';
    B_b(3,2*ndof+1:3*ndof)= XYderivatives(:,1)';

% mass matrix

    mass(indice,indice)=mass(indice,indice)+...
        shapeFunction*shapeFunction'*thickness*...
        rho*gaussWeights(q)*det(Jacob);
    mass(indice+numberNodes,indice+numberNodes)=...
        mass(indice+numberNodes,indice+numberNodes)+...
        shapeFunction*shapeFunction'*I*...
        rho*gaussWeights(q)*det(Jacob);
    mass(indice+2*numberNodes,indice+2*numberNodes)=...
        mass(indice+2*numberNodes,indice+2*numberNodes)+...
        shapeFunction*shapeFunction'*I*...
        rho*gaussWeights(q)*det(Jacob);

    end   % Gauss point
end   % element
```

12.6 Buckling analysis of Mindlin plates

In this section we formulate and implement the buckling analysis of Mindlin plates. After presenting the basic finite element formulation, we present a MATLAB code for buckling analysis of a simply-supported isotropic square plate under uniaxial initial stress.

The strain energy for an initially stressed Mindlin plate, after neglecting terms with third and higher powers in displacement gradients, can be written as [8]

$$U = \frac{1}{2} \int_V \epsilon^T \mathbf{Q} \epsilon dV + \frac{1}{2} \int_V \gamma^T \mathbf{Q}_c \gamma dV + \int_V \left(\sigma^0\right)^T \epsilon^L dV \qquad (12.32)$$

where

$$\epsilon^L = \begin{bmatrix} \frac{1}{2}\left(\left(\frac{\partial u}{\partial x}\right)^2 + \left(\frac{\partial v}{\partial x}\right)^2 + \left(\frac{\partial w}{\partial x}\right)^2\right) \\ \frac{1}{2}\left(\left(\frac{\partial u}{\partial y}\right)^2 + \left(\frac{\partial v}{\partial y}\right)^2 + \left(\frac{\partial w}{\partial y}\right)^2\right) \\ \frac{\partial u}{\partial x}\frac{\partial u}{\partial y} + \frac{\partial v}{\partial x}\frac{\partial v}{\partial y} + \frac{\partial w}{\partial x}\frac{\partial w}{\partial y} \end{bmatrix} \quad (12.33)$$

Integrating over the plate thickness we obtain U as

$$U = \frac{1}{2}\int_A \epsilon^T \mathbf{D}_f \epsilon dA + \frac{1}{2}\int_A \gamma^T \mathbf{D}_c \gamma dA + \frac{1}{2}\int_A \left[\left(\frac{\partial w}{\partial x}\right)\left(\frac{\partial w}{\partial y}\right)\right]\hat{\sigma}^{0T}\begin{bmatrix}\left(\frac{\partial w}{\partial x}\right) \\ \left(\frac{\partial w}{\partial y}\right)\end{bmatrix} h dA$$

$$+ \frac{1}{2}\int_A \left[\left(\frac{\partial \theta_x}{\partial x}\right)\left(\frac{\partial \theta_x}{\partial y}\right)\right]\hat{\sigma}^{0T}\begin{bmatrix}\left(\frac{\partial \theta_x}{\partial x}\right) \\ \left(\frac{\partial \theta_x}{\partial y}\right)\end{bmatrix}\frac{h^3}{12}dA$$

$$+ \frac{1}{2}\int_A \left[\left(\frac{\partial \theta_y}{\partial x}\right)\left(\frac{\partial \theta_y}{\partial y}\right)\right]\hat{\sigma}^{0T}\begin{bmatrix}\left(\frac{\partial \theta_y}{\partial x}\right) \\ \left(\frac{\partial \theta_y}{\partial y}\right)\end{bmatrix}\frac{h^3}{12}dA \quad (12.34)$$

where

$$\hat{\sigma}^0 = \begin{bmatrix} \sigma_x^0 & \tau_{xy}^0 \\ \tau_{xy}^0 & \sigma_y^0 \end{bmatrix} \quad (12.35)$$

The stability problem involves the solution of the eigenproblem

$$[\mathbf{K} - \lambda \mathbf{K_G}]\,\mathbf{a}^i = 0, \quad i = 1, 2, ..., r \quad (12.36)$$

where $\mathbf{K}$ is the global stiffness matrix, $\mathbf{K_G}$ is the geometric matrix and λ is a constant by which the in-plane loads must be multiplied to cause buckling. Vector $\mathbf{a}^i$ is the i-th buckling mode and r is the total number of degrees of freedom. Thus the buckling loads can be found by solving the eigenproblem in (12.36).

The geometric stiffness matrix $\mathbf{K_G}$ may be written as [8].

$$K_G = K_{Gb} + K_{Gs} \quad (12.37)$$

where the bending contribution is given by

$$K_{Gb} = \int_{-1}^{1} \int_{-1}^{1} \mathbf{G}_b^T \hat{\sigma}^0 \mathbf{G}_b h \, det\mathbf{J} d\xi d\eta \tag{12.38}$$

and for a given node i

$$\mathbf{G}_b = \begin{bmatrix} \dfrac{\partial N}{\partial x} & 0 & 0 \\[2mm] \dfrac{\partial N}{\partial y} & 0 & 0 \end{bmatrix} \tag{12.39}$$

The shear contribution is given as

$$K_{Gs} = \int_{-1}^{1} \int_{-1}^{1} \mathbf{G}_{s1}^T \hat{\sigma}^0 \mathbf{G}_{s1} \frac{h^3}{12} det\mathbf{J} d\xi d\eta + \int_{-1}^{1} \int_{-1}^{1} \mathbf{G}_{s2}^T \hat{\sigma}^0 \mathbf{G}_{s2} \frac{h^3}{12} det\mathbf{J} d\xi d\eta \tag{12.40}$$

where for a given node i

$$\mathbf{G}_{s1} = \begin{bmatrix} 0 & \dfrac{\partial N}{\partial x} & 0 \\[2mm] 0 & \dfrac{\partial N}{\partial y} & 0 \end{bmatrix} \tag{12.41}$$

$$\mathbf{G}_{s2} = \begin{bmatrix} 0 & 0 & \dfrac{\partial N}{\partial x} \\[2mm] 0 & 0 & \dfrac{\partial N}{\partial y} \end{bmatrix} \tag{12.42}$$

Although the shear contribution for the geometric stiffness matrix is negligible for thin plates, its effects can be significant for thicker plates.

Table 12.14 summarizes results for simply supported square plates of various thicknesses under uniaxial σ_{xx} initial stress. We consider a 10×10 Q4 mesh (figure 12.7), and compare present the finite element formulation with closed form solution [8]. The schematic geometry, loads and boundary conditions are illustrated in figure 12.8.

In figures 12.9–12.11 the eigenmodes are illustrated, for $a/h = 10$.

The MATLAB code **problem19Buckling.m** computes the problem of a Mindlin plate under compressive, uniaxial load. Note that the code is given as a function in order to be able to incorporate other functions in the same file.

Table 12.14 Buckling factors $k_b = a^2 F_{cr}/\pi^2 D_{f11}$ for a simply supported square plate under uniaxial initial stress ($\nu = 0.3$)

a/h	Closed form solution	Present FE formulation
1,000	4.000	4.0666
20	3.944	3.9930
10	3.786	3.7890
5	3.264	3.1665

Fig. 12.7 Buckling of
Mindlin plate: 10×10 Q4
mesh

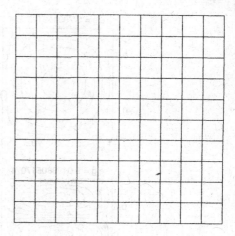

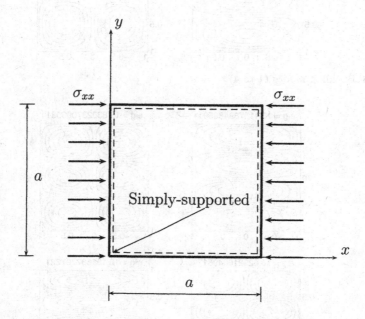

Fig. 12.8 Buckling problem: a Mindlin plate under uniaxial initial stress

This code calls function formGeometricStiffnessMindlinQ4.m for the computation of the geometric stiffness matrix.

```
%...................................................................

% MATLAB codes for Finite Element Analysis
```

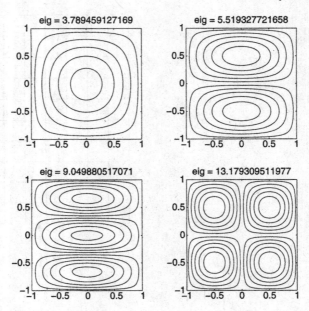

Fig. 12.9 Buckling modes (1 to 4)

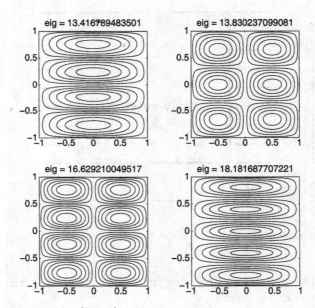

Fig. 12.10 Buckling modes (5 to 8)

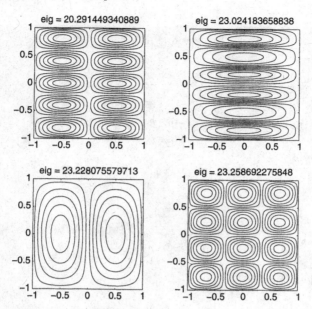

Fig. 12.11 Buckling modes (9 to 12)

```
% problem19Buckling.m
% this function performs buckling analysis
% of Mindlin plates using 3 degrees of freedom per node
% antonio ferreira 2008

clear all
colordef white

% material properties
% modulusOfElasticity  = Young's modulus
% PoissonRatio         = Poisson's ratio

modulusOfElasticity = 10920;  % Young
PoissonRatio = 0.30;  % coef. Poisson

% L: side lenght
L  = 1;

thickness=0.001;
I=thickness^3/12;

% kapa: shear correction factor
```

```
kapa=5/6;

% constitutive matrix
% bending part
C_bending=...
    I*modulusOfElasticity/(1-PoissonRatio^2)*...
    [    1                    PoissonRatio              0 ;
         PoissonRatio         1                         0 ;
         0                    0       (1-PoissonRatio)/2 ];

% shear part
C_shear=...
kapa*thickness*modulusOfElasticity/2/(1+PoissonRatio)*eye(2);

% initial stress matrix
sigmaX=1/thickness;
sigmaXY=0;
sigmaY=0;
sigmaMatrix=[ sigmaX sigmaXY; sigmaXY sigmaY];

% mesh generation ...
% numberElementsX: number of elements in x
% numberElementsY: number of elements in y
numberElementsX=10;
numberElementsY=10;
% number of elements
numberElements=numberElementsX*numberElementsY;
[nodeCoordinates, elementNodes] = ...
    rectangularMesh(L, L, numberElementsX, numberElementsY);
xx=nodeCoordinates(:,1);   yy=nodeCoordinates(:,2);
figure
drawingMesh(nodeCoordinates,elementNodes,'Q4','k-');
axis off

numberNodes=size(xx,1);    % number of nodes
GDof=3*numberNodes;        % total number of DOFs

% stiffness and geometric stiffness matrices
[stiffness]=...
    formStiffnessMatrixMindlinQ4(GDof,numberElements,...
    elementNodes,numberNodes,nodeCoordinates,C_shear,...
    C_bending,thickness,I);
```

```
[geometric]=...
formGeometricStiffnessMindlinQ4(GDof,numberElements,...
elementNodes,numberNodes,nodeCoordinates,sigmaMatrix,thickness);

% Essential boundary conditions
[prescribedDof,activeDof,fixedNodeW]=...
EssentialBC('ssss',GDof,xx,yy,nodeCoordinates,numberNodes);

% buckling analysis ...

% perform eigenproblem
[V1,D1] = eig(stiffness(activeDof,activeDof),...
    geometric(activeDof,activeDof));
D1 = diag(D1);
% drawing eigenmodes
numberOfModes=12;
% sort out eigenvalues
[D1,ii] = sort(D1); ii = ii(1:numberOfModes);
VV = V1(:,ii);
activeDofW=setdiff([1:numberNodes]',[fixedNodeW]);
NNN=size(activeDofW);

% normalize results
disp('D1(1)/pi/pi/C_bending(1,1)')
D1(1)/pi/pi/C_bending(1,1)
D1(1)*pi*pi*C_bending(1,1)

VVV(1:numberNodes,1:numberOfModes)=0;
for i=1:numberOfModes
    VVV(activeDofW,i)=VV(1:NNN,i);
end
%
NN=numberNodes;N=sqrt(NN);
x=linspace(-L,L,numberElementsX+1);
y=linspace(-L,L,numberElementsY+1);

D1=D1/pi/pi/C_bending(1,1);
% drawing Eigenmodes
drawEigenmodes2D(x,y,VVV,NN,N,D1)
```

```
%.................................................................

function [KG]=...
    formGeometricStiffnessMindlinQ4(GDof,numberElements,...
    elementNodes,numberNodes,nodeCoordinates,sigmaMatrix,thickness)

% computation of geometric stiffness
% for Mindlin plate element

% KG : geometric matrix
KG=zeros(GDof);

% Gauss quadrature for bending part
[gaussWeights,gaussLocations]=gaussQuadrature('reduced');

% cycle for element
for e=1:numberElements
    % indice : nodal condofectivities for each element
    % elementDof: element degrees of freedom
    indice=elementNodes(e,:);
    elementDof=[indice indice+numberNodes indice+2*numberNodes];
    ndof=length(indice);

    % cycle for Gauss point
    for q=1:size(gaussWeights,1)
        GaussPoint=gaussLocations(q,:);
        xi=GaussPoint(1);
        eta=GaussPoint(2);

% shape functions and derivatives
        [shapeFunction,naturalDerivatives]=shapeFunctionQ4(xi,eta)

% Jacobian matrix, inverse of Jacobian,
% derivatives w.r.t. x,y
        [Jacob,invJacobian,XYderivatives]=...
            Jacobian(nodeCoordinates(indice,:),naturalDerivatives);

% geometric matrix
        G_b=zeros(2,3*ndof);
        G_b(1,1:ndof)  = XYderivatives(:,1)';
        G_b(2,1:ndof)  = XYderivatives(:,2)';
        KG(elementDof,elementDof)=KG(elementDof,elementDof)+ ...
```

```
        G_b'*sigmaMatrix*thickness*G_b*gaussWeights(q)*det(Jacob);

    end   % Gauss point

  end     % element

% shear stiffness matrix

% Gauss quadrature for shear part
[W,Q]=gaussQuadrature('reduced');

% cycle for element
  for q=1:size(gaussWeights,1)
    GaussPoint=gaussLocations(q,:);
    xi=GaussPoint(1);
    eta=GaussPoint(2);

% shape functions and derivatives
    [shapeFunction,naturalDerivatives]=shapeFunctionQ4(xi,eta)

% Jacobian matrix, inverse of Jacobian,
% derivatives w.r.t. x,y
    [Jacob,invJacobian,XYderivatives]=...
        Jacobian(nodeCoordinates(indice,:),naturalDerivatives);

% Geometric matrix
    G_s1=zeros(2,3*ndof);
    G_s1(1,ndof+1:2*ndof)     = XYderivatives(:,1)';
    G_s1(2,ndof+1:2*ndof)     = XYderivatives(:,2)';
    KG(elementDof,elementDof) =KG(elementDof,elementDof)+ ...
    G_s1'*sigmaMatrix*thickness^3/12*G_s1*...
    gaussWeights(q)*det(Jacob);

    G_s2=zeros(2,3*ndof);
    G_s2(1,2*ndof+1:3*ndof)   = XYderivatives(:,1)';
    G_s2(2,2*ndof+1:3*ndof)   = XYderivatives(:,2)';
    KG(elementDof,elementDof) =KG(elementDof,elementDof)+ ...
    G_s2'*sigmaMatrix*thickness^3/12*G_s2*...
    gaussWeights(q)*det(Jacob);

  end   % gauss point
 end    % element
```

Codes call function drawEigenmodes2D.p. Note that p-codes are not editable.

Chapter 13
Laminated plates

13.1 Introduction

Here we consider a first order shear deformation theory for the static and free vibration analysis of laminated plates. We introduce a computation of the shear correction factor and solve some examples with MATLAB codes.

13.2 Displacement field

In the first order shear deformation theory, displacements are the same as in Mindlin plate theory,

$$u(x, y, z) = u_0(x, y) + z\theta_x(x, y)$$
$$v(x, y, z) = v_0(x, y) + z\theta_y(x, y)$$
$$w(x, y, z) = w_0(x, y) \tag{13.1}$$

13.3 Strains

Strains are obtained by derivation of displacements as

A.J.M. Ferreira, *MATLAB Codes for Finite Element Analysis:*
Solids and Structures, Solid Mechanics and Its Applications 157,
© Springer Science+Business Media B.V. 2009

$$\begin{Bmatrix} \epsilon_{xx} \\ \epsilon_{yy} \\ \gamma_{xy} \\ \gamma_{xz} \\ \gamma_{yz} \end{Bmatrix} = \begin{Bmatrix} \dfrac{\partial u}{\partial x} \\[2mm] \dfrac{\partial v}{\partial y} \\[2mm] \dfrac{\partial u}{\partial y} + \dfrac{\partial v}{\partial x} \\[2mm] \dfrac{\partial u}{\partial z} + \dfrac{\partial w}{\partial x} \\[2mm] \dfrac{\partial v}{\partial z} + \dfrac{\partial w}{\partial y} \end{Bmatrix} \qquad (13.2)$$

or

$$\boldsymbol{\epsilon} = \begin{Bmatrix} \epsilon_{xx} \\ \epsilon_{yy} \\ \gamma_{xy} \end{Bmatrix} = \begin{Bmatrix} \epsilon_{xx}^m \\ \epsilon_{yy}^m \\ \gamma_{xy}^m \end{Bmatrix} + z \begin{Bmatrix} \epsilon_{xx}^f \\ \epsilon_{yy}^f \\ \gamma_{xy}^f \end{Bmatrix} \qquad (13.3)$$

$$\boldsymbol{\gamma} = \begin{Bmatrix} \gamma_{xz} \\ \gamma_{yz} \end{Bmatrix} = \begin{Bmatrix} \gamma_{xz}^{(0)} \\ \gamma_{yz}^{(0)} \end{Bmatrix} \qquad (13.4)$$

where the deformation components are described as

$$\begin{Bmatrix} \epsilon_{xx}^m \\ \epsilon_{yy}^m \\ \gamma_{xy}^m \end{Bmatrix} = \begin{Bmatrix} \dfrac{\partial u_0}{\partial x} \\[2mm] \dfrac{\partial v_0}{\partial y} \\[2mm] \dfrac{\partial u_0}{\partial y} + \dfrac{\partial v_0}{\partial x} \end{Bmatrix} ; \begin{Bmatrix} \epsilon_{xx}^f \\ \epsilon_{yy}^f \\ \gamma_{xy}^f \end{Bmatrix} = \begin{Bmatrix} \dfrac{\partial \theta_x}{\partial x} \\[2mm] \dfrac{\partial \theta_y}{\partial y} \\[2mm] \dfrac{\partial \theta_x}{\partial y} + \dfrac{\partial \theta_y}{\partial x} \end{Bmatrix} \qquad (13.5)$$

$$\begin{Bmatrix} \gamma_{xz}^{(0)} \\ \gamma_{yz}^{(0)} \end{Bmatrix} = \begin{Bmatrix} \dfrac{\partial w_0}{\partial x} + \theta_x \\[2mm] \dfrac{\partial w_0}{\partial y} + \theta_y \end{Bmatrix} \qquad (13.6)$$

13.4 Strain-displacement matrix B

The corresponding strain-displacement matrices **B** are described in detail in the following. The membrane component is given by

$$
\mathbf{B}_m^{(e)} =
\begin{bmatrix}
\dfrac{\partial N}{\partial x} & 0 & 0 & 0 & 0 \\[2ex]
0 & \dfrac{\partial N}{\partial y} & 0 & 0 & 0 \\[2ex]
\dfrac{\partial N}{\partial y} & \dfrac{\partial N}{\partial x} & 0 & 0 & 0
\end{bmatrix}
\tag{13.7}
$$

the bending component

$$
\mathbf{B}_f^{(e)} =
\begin{bmatrix}
0 & 0 & 0 & \dfrac{\partial N}{\partial x} & 0 \\[2ex]
0 & 0 & 0 & 0 & \dfrac{\partial N}{\partial y} \\[2ex]
0 & 0 & 0 & \dfrac{\partial N}{\partial y} & \dfrac{\partial N}{\partial x}
\end{bmatrix}
\tag{13.8}
$$

and the shear component

$$
\mathbf{B}_c^{(e)} =
\begin{bmatrix}
0 & 0 & \dfrac{\partial N}{\partial x} & N & 0 \\[2ex]
0 & 0 & \dfrac{\partial N}{\partial y} & 0 & N
\end{bmatrix}
\tag{13.9}
$$

13.5 Stresses

By assuming a null transverse normal stress σ_z, the stress-strain relations can be set as

$$
\boldsymbol{\sigma} =
\begin{Bmatrix} \sigma_x \\ \sigma_y \\ \tau_{xy} \end{Bmatrix} =
\begin{bmatrix}
\dfrac{E}{1-\nu^2} & \nu\dfrac{E}{1-\nu^2} & 0 \\[2ex]
\nu\dfrac{E}{1-\nu^2} & \dfrac{E}{1-\nu^2} & 0 \\[2ex]
0 & 0 & G
\end{bmatrix}
\begin{Bmatrix} \epsilon_x \\ \epsilon_y \\ \gamma_{xy} \end{Bmatrix} = \mathbf{D}\boldsymbol{\epsilon}
\tag{13.10}
$$

$$
\boldsymbol{\tau} = \begin{Bmatrix} \tau_{xz} \\ \tau_{yz} \end{Bmatrix} =
\begin{bmatrix} K_1 G & 0 \\ 0 & K_2 G \end{bmatrix}
\begin{Bmatrix} \gamma_{xz} \\ \gamma_{yz} \end{Bmatrix} = \mathbf{D}_c \boldsymbol{\gamma}
\tag{13.11}
$$

being K_1, K_2 the shear correction factors in both directions. At each layer interface, the transverse shear continuity must be guaranteed. The equilibrium equation in x direction is written as

$$\frac{\partial \sigma_x}{\partial x} + \frac{\partial \tau_{xy}}{\partial y} + \frac{\partial \tau_{xz}}{\partial z} = 0 \qquad (13.12)$$

Assume cylindrical bending

$$\tau_{xz} = -\int_{-h/2}^{z} \frac{\partial \sigma_x}{\partial x} dz = -\int_{-h/2}^{z} \frac{\partial M_x}{\partial x} \frac{D_1(z)}{R_1} z dz = -\frac{Q_x}{R_1} \int_{-h/2}^{z} D_1(z) z dz = \frac{Q_x}{R_1} g(z)$$
$$(13.13)$$

where

- Q_x is the shear force in xz plane.
- $R_1 = \displaystyle\int_{-h/2}^{h/2} D_1(z) z^2 dz$ represents the plate stiffness in x direction.
- z is the thickness coordinate.
- $g(z) = -\displaystyle\int_{-h/2}^{z} D_1(z) z dz$ represents the shear shape.

Function $g(z)$ which represents the shear stress diagram becomes parabolic for homogeneous sections, $g(z) = [D_1 h^2/8][1 - 4(z/h)^2]$. The strain energy is given by

$$w_s = \int_{-h/2}^{h/2} \frac{\tau_{xz}^2}{G_{13}(z)} dz = \frac{Q_x^2}{R_1^2} \int_{-h/2}^{h/2} \frac{g^2(z)}{G_{13}(z)} dz \qquad (13.14)$$

being $G_{13}(z)$ is the transverse shear modulus in xz plane. For a constant transverse shear deformation the strain is given by

$$\overline{w}_s = \int_{-h/2}^{h/2} \overline{\gamma}_{xz} G_{13}(z) \overline{\gamma}_{xz} dz = \frac{Q_x^2}{h^2 \overline{G}_1^2} h \overline{G}_1 = \frac{Q_x^2}{h \overline{G}_1} \qquad (13.15)$$

where

$$h \overline{G}_1 = \int_{-h/2}^{h/2} G_{13}(z) dz \qquad (13.16)$$

and $\overline{\gamma}_{xz}$ is the mean value for transverse shear strain. It is now possible to obtain the shear correction factor k_1 as

$$k_1 = \frac{\overline{w}_s}{w_s} = \frac{R_1^2}{h \overline{G}_1 \displaystyle\int_{-h/2}^{h/2} g^2(z)/G_{13}(z) dz} \qquad (13.17)$$

To obtain the second shear factor k_2 we proceed in a similar way [21].

13.6 Stiffness matrix

Figure 13:1 illustrates the position of the z coordinates across the thickness direction.

The strain energy is

$$\mathbf{U} = \frac{1}{2} \int_V \epsilon^T \sigma + \gamma^T \tau dV \qquad (13.18)$$

or

$$\mathbf{U} =$$

$$\frac{1}{2} \int_A \mathbf{u}^T \int_z \left[\mathbf{B_m}^T \mathbf{D}^k \mathbf{B_m} + \mathbf{B_m}^T z \mathbf{D}^k \mathbf{B_f} + \mathbf{B_f}^T z \mathbf{D}^k \mathbf{B_m} + \mathbf{B_f}^T z^2 \mathbf{D}^k \mathbf{B_m f} \right] dz \mathbf{u} dA +$$

$$\frac{1}{2} \int_A \mathbf{u}^T \int_z \left[\mathbf{B_c}^T \mathbf{D_c}^k \mathbf{B_c} \right] dz \mathbf{u} dA \quad (13.19)$$

The stiffness matrix is now decomposed into more components, including a membrane-bending coupling part,

$$\mathbf{K}^{(e)} = \mathbf{K}_{mm}^{(e)} + \mathbf{K}_{mf}^{(e)} + \mathbf{K}_{fm}^{(e)} + \mathbf{K}_{ff}^{(e)} + \mathbf{K}_{cc}^{(e)} \qquad (13.20)$$

where $\mathbf{K}_{mm}^{(e)}$ is the membrane part of the stiffness matrix, $\mathbf{K}_{mf}^{(e)}, \mathbf{K}_{fm}^{(e)}$ are the membrane-bending coupling components, $\mathbf{K}_{ff}^{(e)}$ is the bending part, and $\mathbf{K}_{cc}^{(e)}$ is the shear part, detailed as

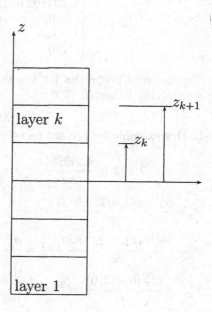

Fig. 13.1 Laminated plate: organization of layers in the thickness direction

$$\mathbf{K}_{mm}^{(e)} = \sum_{k=1}^{nc} \int_A \mathbf{B}_m^T \mathbf{D}_k \mathbf{B}_m \left(z_{k+1} - z_k \right) dA \qquad (13.21)$$

$$\mathbf{K}_{mf}^{(e)} = \sum_{k=1}^{nc} \int_A \mathbf{B}_m^T \mathbf{D}_k \mathbf{B}_f \frac{1}{2} \left(z_{k+1}^2 - z_k^2 \right) dA \qquad (13.22)$$

$$\mathbf{K}_{fm}^{(e)} = \sum_{k=1}^{nc} \int_A \mathbf{B}_f^T \mathbf{D}_k \mathbf{B}_m \frac{1}{2} \left(z_{k+1}^2 - z_k^2 \right) dA \qquad (13.23)$$

$$\mathbf{K}_{ff}^{(e)} = \sum_{k=1}^{nc} \int_A \mathbf{B}_f^T \mathbf{D}_k \mathbf{B}_f \frac{1}{3} \left(z_{k+1}^3 - z_k^3 \right) dA \qquad (13.24)$$

$$\mathbf{K}_{cc}^{(e)} = \sum_{k=1}^{nc} \int_A \mathbf{B}_c^T \mathbf{D}_k \mathbf{B}_c \left(z_{k+1} - z_k \right) dA \qquad (13.25)$$

where nc denotes the number of layers across the thickness direction.

13.7 Numerical example

We analyse a laminated sandwich three-layer plate, simply-supported on all sides, under uniform pressure. This is known as the Srinivas problem [22], with the following core properties:

$$\overline{Q}_{core} = \begin{bmatrix} 0.999781 & 0.231192 & 0 & 0 & 0 \\ 0.231192 & 0.524886 & 0 & 0 & 0 \\ 0 & 0 & 0.262931 & 0 & 0 \\ 0 & 0 & 0 & 0.266810 & 0 \\ 0 & 0 & 0 & 0 & 0.159914 \end{bmatrix}$$

The material properties for the skins are obtained from those of the core and a multiplying factor R:

$$\overline{Q}_{skin} = R\overline{Q}_{core}$$

In this example we present non-dimensional results

$$\overline{w} = w(a/2, a/2, 0)\frac{0.999781}{hq}$$

$$\overline{\sigma}_x^1 = \frac{\sigma_x^{(1)}(a/2, a/2, -h/2)}{q}; \overline{\sigma}_x^2 = \frac{\sigma_x^{(1)}(a/2, a/2, -2h/5)}{q}; \quad \overline{\sigma}_x^3 = \frac{\sigma_x^{(2)}(a/2, a/2, -2h/5)}{q}$$

$$\overline{\sigma}_y^1 = \frac{\sigma_y^{(1)}(a/2, a/2, -h/2)}{q} \overline{\sigma}_y^2 = \frac{\sigma_y^{(1)}(a/2, a/2, -2h/5)}{q}; \quad \overline{\sigma}_y^3 = \frac{\sigma_y^{(2)}(a/2, a/2, -2h/5)}{q}$$

$$\overline{\tau}_{xz}^1 = \frac{\tau_{xz}^{(2)}(0, a/2, 0)}{q}; \quad \overline{\tau}_{xz}^2 = \frac{\tau_{xz}^{(2)}(0, a/2, -2h/5)}{q}$$

For various values of R, we compare results with third-order theory of Pandya [23], and various finite element and meshless results by the author. Results are quite good (as illustrated in tables 13.1 to 13.3), with the exception of the transverse shear stresses that should be further corrected [21],

$$\tau_{xz}^{cor} = \bar{G}_{13}\gamma_{xz}\frac{g(z)}{\bar{g}} \tag{13.26}$$

where

$$\bar{g} = -\int_{-h/2}^{h/2} g(z)dz \tag{13.27}$$

A viable alternative for the computation of the transverse shear stresses is to solve the equilibrium equations.

Table 13.1 Square sandwich plate under uniform pressure – $R = 5$

Method	$\overline{w}$	$\overline{\sigma}_x^1$	$\overline{\sigma}_x^2$	$\overline{\sigma}_x^3$	$\overline{\tau}_{xz}^1$	$\overline{\tau}_{xz}^2$
HSDT [23]	256.13	62.38	46.91	9.382	3.089	2.566
FSDT [23]	236.10	61.87	49.50	9.899	3.313	2.444
CLT	216.94	61.141	48.623	9.783	4.5899	3.386
Ferreira [24]	258.74	59.21	45.61	9.122	3.593	3.593
Ferreira (N = 15) [25]	257.38	58.725	46.980	9.396	3.848	2.839
Analytical [22]	258.97	60.353	46.623	9.340	4.3641	3.2675
HSDT [26] (N = 11)	253.6710	59.6447	46.4292	9.2858	3.8449	1.9650
HSDT [26] (N = 15)	256.2387	60.1834	46.8581	9.3716	4.2768	2.2227
HSDT [26] (N = 21)	257.1100	60.3660	47.0028	9.4006	4.5481	2.3910
present (4 × 4 elements Q4)	260.0321	54.6108	43.6887	8.7377	2.3922	11.9608
present (10 × 10 elements Q4)	259.3004	58.4403	46.7523	9.3505	2.9841	14.9207
present (20 × 20 elements Q4)	259.2797	58.9507	47.1606	9.4321	3.1980	15.9902

Table 13.2 Square sandwich plate under uniform pressure – $R = 10$

Method	$\overline{w}$	$\overline{\sigma}_x^1$	$\overline{\sigma}_x^2$	$\overline{\sigma}_x^3$	$\overline{\tau}_{xz}^1$	$\overline{\tau}_{xz}^2$
HSDT [23]	152.33	64.65	51.31	5.131	3.147	2.587
FSDT [23]	131.095	67.80	54.24	4.424	3.152	2.676
CLT	118.87	65.332	48.857	5.356	4.3666	3.7075
Ferreira [24]	159.402	64.16	47.72	4.772	3.518	3.518
Ferreira (N = 15) [25]	158.55	62.723	50.16	5.01	3.596	3.053
Analytical [22]	159.38	65.332	48.857	4.903	4.0959	3.5154
HSDT [26] (N = 11)	153.0084	64.7415	49.4716	4.9472	2.7780	1.8207
HSDT [26] (N = 15)	154.2490	65.2223	49.8488	4.9849	3.1925	2.1360
HSDT [26] (N = 21)	154.6581	65.3809	49.9729	4.9973	3.5280	2.3984
present (4 × 4 elements Q4)	162.2395	58.1236	46.4989	4.6499	1.5126	15.1261
present (10 × 10 elements Q4)	159.9120	62.3765	49.9012	4.9901	1.8995	18.9954
present (20 × 20 elements Q4)	159.6820	62.9474	50.3580	5.0358	2.0371	20.3713

Table 13.3 Square sandwich plate under uniform pressure – $R = 15$

Method	$\overline{w}$	$\overline{\sigma}_x^1$	$\overline{\sigma}_x^2$	$\overline{\sigma}_x^3$	$\overline{\tau}_{xz}^1$	$\overline{\tau}_{xz}^2$
HSDT [23]	110.43	66.62	51.97	3.465	3.035	2.691
FSDT [23]	90.85	70.04	56.03	3.753	3.091	2.764
CLT	81.768	69.135	55.308	3.687	4.2825	3.8287
Ferreira [24]	121.821	65.650	47.09	3.140	3.466	3.466
Ferreira (N = 15) [25]	121.184	63.214	50.571	3.371	3.466	3.099
Analytical [22]	121.72	66.787	48.299	3.238	3.9638	3.5768
HSDT [26] (N = 11)	113.5941	66.3646	49.8957	3.3264	2.1686	1.5578
HSDT [26] (N = 15)	114.3874	66.7830	50.2175	3.3478	2.6115	1.9271
HSDT [26] (N = 21)	114.6442	66.9196	50.3230	3.3549	3.0213	2.2750
present (4 × 4 elements Q4)	125.2176	58.4574	46.7659	3.1177	1.0975	16.4621
present (10 × 10 elements Q4)	122.3318	62.8602	50.2881	3.3525	1.3857	20.7849
present (20 × 20 elements Q4)	122.0283	63.4574	50.7659	3.3844	1.4872	22.3084

Code problem20.m solves this problem.

```
%...............................................

% MATLAB codes for Finite Element Analysis
% problem20.m
% laminated plate: Srinivas problem:
% S. Srinivas, A refined analysis of composite laminates,
% J. Sound and Vibration, 30 (1973),495--507.

% antonio ferreira 2008

% clear memory
clear all;colordef white;clf

% materials
thickness=0.1;

% load
P = -1;

%Mesh generation
L  = 1;
numberElementsX=10;
numberElementsY=10;
numberElements=numberElementsX*numberElementsY;
```

```
[nodeCoordinates, elementNodes] = ...
    rectangularMesh(L,L,numberElementsX,numberElementsY);
xx=nodeCoordinates(:,1);
yy=nodeCoordinates(:,2);
drawingMesh(nodeCoordinates,elementNodes,'Q4','k-');
axis off
numberNodes=size(xx,1);

% GDof: global number of degrees of freedom
GDof=5*numberNodes;

% computation of the system stiffness matrix
% the shear correction factors are automatically
% calculted for any laminate

[AMatrix,BMatrix,DMatrix,SMatrix,qbarra]=srinivasMaterial
    (thickness);

stiffness=formStiffnessMatrixMindlinQ45laminated5dof...
    (GDof,numberElements,...
    elementNodes,numberNodes,nodeCoordinates,AMatrix,...
    BMatrix,DMatrix,SMatrix);

% computation of the system force vector
[force]=...
    formForceVectorMindlinQ45dof(GDof,numberElements,...
    elementNodes,numberNodes,nodeCoordinates,P);

% boundary conditions
[prescribedDof,activeDof,fixedNodeW]=...
    EssentialBC5dof('ssss',GDof,xx,yy,nodeCoordinates,
        numberNodes);

% solution
U=solution(GDof,prescribedDof,stiffness,force);

% drawing deformed shape and normalize results
% to compare with Srinivas
ws=1:numberNodes;
disp('maximum displacement')
abs(min(U(ws))*0.999781/thickness)
figure (1)
```

```
plot3(xx,yy,U(ws),'.')

% stress computation (Srinivas only)
disp('stress computation (Srinivas only)')
SrinivasStress(GDof,numberElements,...
    elementNodes,numberNodes,nodeCoordinates,qbarra,U,thickness);
```

The computation of the material constitutive matrices is made in function srini-
vasMaterial.m,

```
function [AMatrix,BMatrix,DMatrix,SMatrix,qbarra]=...
    srinivasMaterial(thickness)

%%% SRINIVAS EXAMPLE

dd=zeros(2);d=zeros(3);
% multiplying factor for skins
rf=15;
% plate thickness
h=thickness;
% matrix [D]
% in-plane
dmat=[0.999781 0.231192 0 ;0.231192 0.524886 0;0 0 0.262931];
% shear
dm=[0.26681 0; 0 0.159914];
% nc: number of layers
nc=3;
% layers angles
ttt=0;ttt1=0; th(1)=ttt;th(2)=ttt1;th(3)=ttt1;
% coordinates - z1 (upper) and - z2 (lower) for each layer
z1=[-2*h/5 2*h/5 h/2];
z2=[-h/2 -2*h/5 2*h/5;];
% thickness for each layer
thick(1:nc)=z1(1:nc)-z2(1:nc);
% coefe: shear correction factors (k1 and k2)
coefe(1:2)=0.0;gbarf(1:2)=0.0;rfact(1:2)=0.0;
sumla(1:2)=0.0;trlow(1:2)=0.0;upter(1:2)=0.0;
% middle axis position (bending)
        dsumm=0.0;
        for ilayr=1:nc
            dzeta=z1(ilayr)-z2(ilayr);
            zheig=dsumm+dzeta/2.0;
```

```
                dindx(1)=rf*dmat(1,1);dindx(2)=dmat(2,2);
                upter(1:2)=upter(1:2)+dindx(1:2)*zheig*dzeta;
                trlow(1:2)=trlow(1:2)+dindx(1:2)*dzeta;

            dsumm = dsumm+dzeta;
        end

    zeta2(1:2)=-upter(1:2)./trlow(1:2);

% shear correction factors.

        for  ilayr=1:nc
            diff1=z1(ilayr)-z2(ilayr);
            d1=rf*dmat(1,1);
            d2=rf*dmat(2,2);
            d3=rf*dm(1,1);
            d4=rf*dm(2,2);
            if(ilayr==2)
                d1=dmat(1,1);
                d3=dm(1,1);
                d4=dm(2,2);
                d2=dmat(2,2);
            end
        index=10;
        for i=1:2
            zeta1(i)=zeta2(i);
            zeta2(i)=zeta1(i)+diff1;
            diff2(i)=zeta2(i)^2-zeta1(i)^2;
            diff3(i)=zeta2(i)^3-zeta1(i)^3;
            diff5(i)=zeta2(i)^5-zeta1(i)^5;

            dindx=[d1;d2];
            gindx=[d3;d4];

            gbarf(i)=gbarf(i)+gindx(i)*diff1/2.0;
            rfact(i)=rfact(i)+dindx(i)*diff3(i)/3.0;

            term1   = sumla(i)*sumla(i)*diff1;
            term2   = dindx(i)*(zeta1(i)^4)*diff1/4.0;
            term3   = dindx(i)*diff5(i)/20.0;
            term4   =-dindx(i)*zeta1(i)*zeta1(i)*diff3(i)/6.0;
            term5   = sumla(i)*zeta1(i)*zeta1(i)*diff1;
```

```
            term6    =-sumla(i)*diff3(i)/3.0;
            coefe(i)= coefe(i)+(term1+dindx(i)*...
                (term2+term3+term4+term5+term6))/gindx(i);
            index    = index+1;
            sumla(i)= sumla(i)-dindx(i)*diff2(i)/2.0;
        end
      end

    coefe(1:2)=rfact(1:2).*rfact(1:2)./(2.0*gbarf(1:2).
        *coefe(1:2));
    kapa=coefe(1);

% constitutive matrice, membrane, bending and shear
a11=0;a22=0;a12=0;a33=0;
for i=1:nc
theta=th(i);
q11=rf*dmat(1,1);q12=rf*dmat(1,2);q22=rf*dmat(2,2);q33=rf
    *dmat(3,3);
cs=cos(theta);ss=sin(theta);ss11=rf*dm(1,1)*kapa;ss22=rf*dm(2,2)
    *kapa;
if i==2
    q11=dmat(1,1);q12=dmat(1,2);q22=dmat(2,2);q33=dmat(3,3);
    cs=cos(theta);ss=sin(theta);
    ss11=dm(1,1)*kapa;ss22=dm(2,2)*kapa;
end
dd(1,1)=dd(1,1)+(ss11*cos(theta)^2+ss22*sin(theta)^2)
    *(z1(i)-z2(i));
dd(2,2)=dd(2,2)+(ss11*sin(theta)^2+ss22*cos(theta)^2)
    *(z1(i)-z2(i));
d(1,1)=d(1,1)+(q11*cs^4+2*(q12+2*q33)*ss*ss*cs*cs+...
    q22*ss^4)*(z1(i)^3-z2(i)^3)/3;
d(2,2)=d(2,2)+(q11*ss^4+2*(q12+2*q33)*ss*ss*cs*cs+...
    q22*cs^4)*(z1(i)^3-z2(i)^3)/3;
d(1,2)=d(1,2)+((q11+q22-4*q33)*ss*ss*cs*cs+...
    q12*(ss^4+cs^4))*(z1(i)^3-z2(i)^3)/3;
d(3,3)=d(3,3)+((q11+q22-2*q12-2*q33)*ss*ss*cs*cs+...
    q33*(ss^4+cs^4))*(z1(i)^3-z2(i)^3)/3;
a11=a11+q11*thick(i);
a22=a22+q22*thick(i);
a33=a22+q33*thick(i);
a12=a12+q12*thick(i);

qbarra(1,1,i)=q11;
```

```
qbarra(1,2,i)=q12;
qbarra(2,2,i)=q22;
qbarra(3,3,i)=q33;
qbarra(4,4,i)=ss11;
qbarra(5,5,i)=ss22;

end %nc

A44=dd(2,2);
A55=dd(1,1);
D11=d(1,1);
D12=d(1,2);
D22=d(2,2);
D66=d(3,3);
A11=a11;
A12=a12;
A66=a33;
A22=a22;

AMatrix=[A11,A12,0;A12,A22,0;0,0,A66];
%srinivas case (symmetric)
BMatrix=zeros(3);
%BMatrix=[B11,B12,0;B12,B22,0;0,0,B66]
DMatrix=[D11,D12,0;D12,D22,0;0,0,D66];
SMatrix=[A44,0;0,A55];
```

Because this plate element has five degrees of freedom, instead of three degrees of freedom as in Mindlin plates, some changes were introduced and new functions are needed. Function formStiffnessMatrixMindlinQ45laminated5dof.m computes the stiffness matrix for the Q4 Mindlin plate with five DOFs.

```
%.................................................................

function [K]=...
    formStiffnessMatrixMindlinQ45laminated5dof(GDof,
        numberElements,...
    elementNodes,numberNodes,nodeCoordinates,AMatrix,...
    BMatrix,DMatrix,SMatrix)

% computation of stiffness matrix
% for Mindlin plate element
```

```
% K : stiffness matrix

K=zeros(GDof);

% Gauss quadrature for bending part
[gaussWeights,gaussLocations]=gaussQuadrature('complete');

% cycle for element
for e=1:numberElements
  % indice : nodal condofectivities for each element
  % elementDof: element degrees of freedom
  indice=elementNodes(e,:);
  elementDof=[ indice indice+numberNodes indice+2*numberNodes ...
           indice+3*numberNodes indice+4*numberNodes];
  ndof=length(indice);

  % cycle for Gauss point
  for q=1:size(gaussWeights,1)
    GaussPoint=gaussLocations(q,:);
    xi=GaussPoint(1);
    eta=GaussPoint(2);

% shape functions and derivatives
    [shapeFunction,naturalDerivatives]=shapeFunctionQ4(xi,eta);

% Jacobian matrix, inverse of Jacobian,
% derivatives w.r.t. x,y
    [Jacob,invJacobian,XYderivatives]=...
        Jacobian(nodeCoordinates(indice,:),naturalDerivatives);

% [B] matrix bending
    B_b=zeros(3,5*ndof);
    B_b(1,ndof+1:2*ndof)          = XYderivatives(:,1)';
    B_b(2,2*ndof+1:3*ndof)        = XYderivatives(:,2)';
    B_b(3,ndof+1:2*ndof)          = XYderivatives(:,2)';
    B_b(3,2*ndof+1:3*ndof)        = XYderivatives(:,1)';
% [B] matrix membrane
    B_m=zeros(3,5*ndof);
    B_m(1,3*ndof+1:4*ndof)        = XYderivatives(:,1)';
    B_m(2,4*ndof+1:5*ndof)        = XYderivatives(:,2)';
    B_m(3,3*ndof+1:4*ndof)        = XYderivatives(:,2)';
    B_m(3,4*ndof+1:5*ndof)        = XYderivatives(:,1)';
```

```
% stiffness matrix

% ... bending-bending
    K(elementDof,elementDof)=K(elementDof,elementDof)+...
                    B_b'*DMatrix*B_b*gaussWeights(q)*det(Jacob);
% ... membrane-membrane
    K(elementDof,elementDof)=K(elementDof,elementDof)+...
                    B_m'*AMatrix*B_m*gaussWeights(q)*det(Jacob);
% ... membrane-bending
    K(elementDof,elementDof)=K(elementDof,elementDof)+...
                    B_m'*BMatrix*B_b*gaussWeights(q)*det(Jacob);
% ... bending-membrane
    K(elementDof,elementDof)=K(elementDof,elementDof)+...
                    B_b'*BMatrix*B_m*gaussWeights(q)*det(Jacob);

    end  % Gauss point
end    % element

% shear stiffness matrix

% Gauss quadrature for shear part
[gaussWeights,gaussLocations]=gaussQuadrature('reduced');

% cycle for element
% cycle for element
for e=1:numberElements
  % indice : nodal condofectivities for each element
  % elementDof: element degrees of freedom
  indice=elementNodes(e,:);
  elementDof=[ indice indice+numberNodes indice+2*numberNodes ...
            indice+3*numberNodes indice+4*numberNodes];
  ndof=length(indice);

  % cycle for Gauss point
  for q=1:size(gaussWeights,1)
    GaussPoint=gaussLocations(q,:);
    xi=GaussPoint(1);
    eta=GaussPoint(2);

% shape functions and derivatives
    [shapeFunction,naturalDerivatives]=shapeFunctionQ4(xi,eta);
```

```
% Jacobian matrix, inverse of Jacobian,
% derivatives w.r.t. x,y
    [Jacob,invJacobian,XYderivatives]=...
        Jacobian(nodeCoordinates(indice,:),naturalDerivatives);

% [B] matrix shear
    B_s=zeros(2,5*ndof);
    B_s(1,1:ndof)       = XYderivatives(:,1)';
    B_s(2,1:ndof)       = XYderivatives(:,2)';
    B_s(1,ndof+1:2*ndof) = shapeFunction;
    B_s(2,2*ndof+1:3*ndof)= shapeFunction;

% stiffness matrix shear
    K(elementDof,elementDof)=K(elementDof,elementDof)+...
        B_s'*SMatrix  *B_s*gaussWeights(q)*det(Jacob);
  end  % gauss point
end    % element
```

Function formMassMatrixMindlinQ4laminated5dof.m computes the corresponding mass matrix.

```
%..........................................................

function [M]=...
    formMassMatrixMindlinQ4laminated5dof(GDof,numberElements,...
    elementNodes,numberNodes,nodeCoordinates,rho,thickness,I)

% computation of mass matrix
% for Mindlin plate element

M=zeros(GDof);

% Gauss quadrature for bending part
[gaussWeights,gaussLocations]=gaussQuadrature('complete');

% cycle for element
for e=1:numberElements
  % indice : nodal condofectivities for each element
  indice=elementNodes(e,:);
  ndof=length(indice);
```

```
  % cycle for Gauss point
  for q=1:size(gaussWeights,1)
    GaussPoint=gaussLocations(q,:);
    xi=GaussPoint(1);
    eta=GaussPoint(2);

% shape functions and derivatives
    [shapeFunction,naturalDerivatives]=shapeFunctionQ4(xi,eta);

% Jacobian matrix, inverse of Jacobian,
% derivatives w.r.t. x,y
    [Jacob,invJacobian,XYderivatives]=...
        Jacobian(nodeCoordinates(indice,:),naturalDerivatives);

% [B] matrix bending
    B_b=zeros(3,5*ndof);
    B_b(1,ndof+1:2*ndof)        = XYderivatives(:,1)';
    B_b(2,2*ndof+1:3*ndof)      = XYderivatives(:,2)';
    B_b(3,ndof+1:2*ndof)        = XYderivatives(:,2)';
    B_b(3,2*ndof+1:3*ndof)      = XYderivatives(:,1)';
% [B] matrix membrane
    B_m=zeros(3,5*ndof);
    B_m(1,3*ndof+1:4*ndof)      = XYderivatives(:,1)';
    B_m(2,4*ndof+1:5*ndof)      = XYderivatives(:,2)';
    B_m(3,3*ndof+1:4*ndof)      = XYderivatives(:,2)';
    B_m(3,4*ndof+1:5*ndof)      = XYderivatives(:,1)';

% mass matrix
    M(indice,indice)=M(indice,indice)+...
        shapeFunction*shapeFunction'*thickness*rho*...
        gaussWeights(q)*det(Jacob);

    M(indice+numberNodes,indice+numberNodes)=...
    M(indice+numberNodes,indice+numberNodes)+...
        shapeFunction*shapeFunction'*I*rho*...
        gaussWeights(q)*det(Jacob);

    M(indice+2*numberNodes,indice+2*numberNodes)=...
        M(indice+2*numberNodes,indice+2*numberNodes)+...
        shapeFunction*shapeFunction'*I*rho*...
        gaussWeights(q)*det(Jacob);

    M(indice+3*numberNodes,indice+3*numberNodes)=...
```

```
        M(indice+3*numberNodes,indice+3*numberNodes)+...
        shapeFunction*shapeFunction'*thickness*rho*...
        gaussWeights(q)*det(Jacob);

    M(indice+4*numberNodes,indice+4*numberNodes)=...
        M(indice+4*numberNodes,indice+4*numberNodes)+...
        shapeFunction*shapeFunction'*thickness*rho*...
        gaussWeights(q)*det(Jacob);
    end   % Gauss point
end      % element
```

Function formForceVectorMindlinQ45dof.m computes the corresponding force vector.

```
%.................................................................

function [force]=...
    formForceVectorMindlinQ45dof(GDof,numberElements,...
    elementNodes,numberNodes,nodeCoordinates,P)

% computation of force vector
% for Mindlin plate element

% force : force vector
force=zeros(GDof,1);

% Gauss quadrature for bending part
[gaussWeights,gaussLocations]=gaussQuadrature('reduced');

% cycle for element
for e=1:numberElements
  % indice : nodal connectivities for each element
  indice=elementNodes(e,:);

  % cycle for Gauss point
  for q=1:size(gaussWeights,1)
    GaussPoint=gaussLocations(q,:);
    GaussWeight=gaussWeights(q);
    xi=GaussPoint(1);
    eta=GaussPoint(2);
```

```
% shape functions and derivatives
    [shapeFunction,naturalDerivatives]=shapeFunctionQ4(xi,eta);

% Jacobian matrix, inverse of Jacobian,
% derivatives w.r.t. x,y
    [Jacob,invJacobian,XYderivatives]=...
        Jacobian(nodeCoordinates(indice,:),naturalDerivatives);

% force vector
    force(indice)=force(indice)+shapeFunction*P*...
        det(Jacob)*GaussWeight;
  end  % Gauss point

end    % element
```

Function EssentialBC5dof.m computes the prescribed degrees of freedom in vector form.

```
%................................................................

function [prescribedDof,activeDof,fixedNodeW]=...
    EssentialBC5dof(typeBC,GDof,xx,yy,nodeCoordinates,numberNodes)

% essentialBoundary conditions for recatngular plates (5Dof)
switch typeBC
    case 'ssss'
fixedNodeW =find(xx==max(nodeCoordinates(:,1))|...
                 xx==min(nodeCoordinates(:,1))|...
                 yy==min(nodeCoordinates(:,2))|...
                 yy==max(nodeCoordinates(:,2)));

fixedNodeTX =find(yy==max(nodeCoordinates(:,2))|...
    yy==min(nodeCoordinates(:,2)));
fixedNodeTY =find(xx==max(nodeCoordinates(:,1))|...
    xx==min(nodeCoordinates(:,1)));
fixedNodeU =find(xx==min(nodeCoordinates(:,1)));
fixedNodeV =find(yy==min(nodeCoordinates(:,2)));
    case 'cccc'

fixedNodeW =find(xx==max(nodeCoordinates(:,1))|...
```

```
                          xx==min(nodeCoordinates(:,1))|...
                          yy==min(nodeCoordinates(:,2))|...
                          yy==max(nodeCoordinates(:,2)));
    fixedNodeTX =fixedNodeW;
    fixedNodeTY =fixedNodeTX;
    fixedNodeU =fixedNodeTX;
    fixedNodeV =fixedNodeTX;

end

prescribedDof=[fixedNodeW;fixedNodeTX+numberNodes;...
        fixedNodeTY+2*numberNodes;...
        fixedNodeU+3*numberNodes;fixedNodeV+4*numberNodes];
activeDof=setdiff([1:GDof]',[prescribedDof]);
```

For the Srinivas example, stresses are calculated in function SrinivasStress.m.

```
%..............................................................

function SrinivasStress(GDof,numberElements,...
    elementNodes,numberNodes,nodeCoordinates,qbarra,U,h)

% computes normal and shear stresses forSrinivas case
% note that transverse shear stresses are not corrected

% normal stresses in each layer
  stress_layer1=zeros(numberElements,4,3);
  stress_layer2=zeros(numberElements,4,3);
  stress_layer3=zeros(numberElements,4,3);

% Gauss quadrature for bending part
[gaussWeights,gaussLocations]=gaussQuadrature('complete');

% cycle for element
% cycle for element
for e=1:numberElements
  % indice : nodal connectivities for each element
  % indiceB: element degrees of freedom
  indice=elementNodes(e,:);
  indiceB=[ indice indice+numberNodes indice+2*numberNodes ...
            indice+3*numberNodes indice+4*numberNodes];
```

```
    nn=length(indice);

  % cycle for Gauss point
  for q=1:size(gaussWeights,1)
    pt=gaussLocations(q,:);
    wt=gaussWeights(q);
    xi=pt(1);
    eta=pt(2);

% shape functions and derivatives
    [shapeFunction,naturalDerivatives]=shapeFunctionQ4(xi,eta);

% Jacobian matrix, inverse of Jacobian,
% derivatives w.r.t. x,y
    [Jacob,invJacobian,XYderivatives]=...
        Jacobian(nodeCoordinates(indice,:),naturalDerivatives);

% [B] matrix bending
    B_b=zeros(3,5*nn);
    B_b(1,nn+1:2*nn)        = XYderivatives(:,1)';
    B_b(2,2*nn+1:3*nn)      = XYderivatives(:,2)';
    B_b(3,nn+1:2*nn)        = XYderivatives(:,2)';
    B_b(3,2*nn+1:3*nn)      = XYderivatives(:,1)';
% [B] matrix membrane
    B_m=zeros(3,5*nn);
    B_m(1,3*nn+1:4*nn)      = XYderivatives(:,1)';
    B_m(2,4*nn+1:5*nn)      = XYderivatives(:,2)';
    B_m(3,3*nn+1:4*nn)      = XYderivatives(:,2)';
    B_m(3,4*nn+1:5*nn)      = XYderivatives(:,1)';

% stresses
    stress_layer1(e,q,:)=...
        2*h/5*qbarra(1:3,1:3,2)*B_b*U(indiceB)+...
        qbarra(1:3,1:3,2)*B_m*U(indiceB);
    stress_layer2(e,q,:)=...
        2*h/5*qbarra(1:3,1:3,3)*B_b*U(indiceB)+...
        qbarra(1:3,1:3,3)*B_m*U(indiceB);
    stress_layer3(e,q,:)=...
        h/2*qbarra(1:3,1:3,3)*B_b*U(indiceB)+...
        qbarra(1:3,1:3,3)*B_m*U(indiceB);

    end  % Gauss point
end     % element
```

```
% shear stresses in each layer
% by constitutive equations

    shear_layer1=zeros(numberElements,1,2);
    shear_layer2=zeros(numberElements,1,2);
    shear_layer3=zeros(numberElements,1,2);


% Gauss quadrature for shear part
[gaussWeights,gaussLocations]=gaussQuadrature('reduced');

% cycle for element
% cycle for element
for e=1:numberElements
  % indice : nodal connectivities for each element
  % indiceB: element degrees of freedom
  indice=elementNodes(e,:);
  indiceB=[ indice indice+numberNodes indice+2*numberNodes ...
            indice+3*numberNodes indice+4*numberNodes];
  nn=length(indice);

  % cycle for Gauss point
  for q=1:size(gaussWeights,1)
    pt=gaussLocations(q,:);
    wt=gaussWeights(q);
    xi=pt(1);
    eta=pt(2);

% shape functions and derivatives
    [shapeFunction,naturalDerivatives]=shapeFunctionQ4(xi,eta);

% Jacobian matrix, inverse of Jacobian,
% derivatives w.r.t. x,y
    [Jacob,invJacobian,XYderivatives]=...
        Jacobian(nodeCoordinates(indice,:),naturalDerivatives);

% [B] matrix shear
    B_s=zeros(2,5*nn);
    B_s(1,1:nn)        = XYderivatives(:,1)';
    B_s(2,1:nn)        = XYderivatives(:,2)';
    B_s(1,nn+1:2*nn)   = shapeFunction;
    B_s(2,2*nn+1:3*nn) = shapeFunction;
```

```
      shear_layer1(e,q,:)=qbarra(4:5,4:5,1)*B_s*U(indiceB);
      shear_layer2(e,q,:)=qbarra(4:5,4:5,2)*B_s*U(indiceB);

   end % gauss point
 end      % element

% normalized stresses, look for table in the book
format
[ abs(min(stress_layer3(:,3,1))),...
  abs(min(stress_layer2(:,3,1))), ...
  abs(min(stress_layer1(:,3,1))),...
  max(shear_layer2(:,:,1)),...
  max(shear_layer1(:,:,1))]
```

13.8 Free vibrations of laminated plates

The free vibration problem of laminated plates follows the same procedure as for Mindlin plates. The stiffness matrix is as previously computed and the mass matrix is obtained in a similar way.

We consider various stacking sequences, boundary conditions and thickness-to-side ratios. We consider both square and rectangular plates. In order to compare with other sources, eigenvalues are expressed in terms of the non-dimensional frequency parameter $\bar{\omega}$, defined as

$$\bar{\omega} = (\omega b^2/\pi^2)\sqrt{\frac{\rho h}{D_0}},$$

where

$$D_0 = E_{22}h^3/12(1 - \nu_{12}\nu_{21})$$

Also, a constant shear correction factor $k = \pi^2/12$ is used for all computations.

The examples considered here are limited to thick symmetric cross-ply laminates with layers of equal thickness. The material properties for all layers of the laminates are identical:

$$E_{11}/E_{22} = 40; G_{23} = 0.5E_{22}; G_{13} = G_{12} = 0.6E_{22}; \nu_{12} = 0.25; \nu_{21} = 0.00625$$

We consider SSSS (simply supported on all sides) and CCCC (clamped on all sides) boundary conditions, for their practical interest.

The convergence study of frequency parameters $\bar{\omega} = (\omega b^2/\pi^2)\sqrt{(\rho h/D_0)}$ for three-ply $(0°/90°/0°)$ simply supported SSSS rectangular laminates is performed

Table 13.4 Convergence study of frequency parameters $\bar{\omega} = (\omega b^2 / \pi^2)\sqrt{(\rho h/D_0)}$ for three-ply $(0°/90°/0°)$ simply supported SSSS rectangular laminates

			Modes					
a/b	t/b	Mesh	1	2	3	4	5	6
1	0.001	5×5 Q4 elements	6.9607	10.7831	25.1919	30.8932	32.5750	40.8649
		10×10	6.7066	9.7430	17.8158	26.3878	27.8395	32.3408
		20×120	6.6454	9.5190	16.5801	25.4234	26.8237	27.9061
		30×30	6.6342	9.4789	16.3696	25.2510	26.6419	27.1999
	Liew [27]		6.6252	9.4470	16.2051	25.1146	26.4982	26.6572
	0.20	5×5 Q4 elements	3.5913	6.2812	7.6261	8.8475	11.3313	12.1324
		10×10	3.5479	5.8947	7.3163	8.6545	9.7538	11.2835
		20×20	3.5367	5.8036	7.2366	8.5856	9.3768	10.9971
		30×30	3.5346	5.7870	7.2218	8.5721	9.3087	10.9441
	Liew [27]		3.5939	5.7691	7.3972	8.6876	9.1451	11.2080
2	0.001	5×5 Q4 elements	2.4798	8.0538	8.1040	11.5816	23.5944	23.7622
		10×10	2.3905	6.9399	6.9817	9.9192	15.9748	16.0852
		20×20	2.3689	6.7016	6.7415	9.5617	14.6808	14.7815
		30×30	2.3650	6.6590	6.6986	9.4977	14.4598	14.5588
	Liew [27]		2.3618	6.6252	6.6845	9.4470	14.2869	16.3846
	0.20	5×5 Q4 elements	2.0006	3.7932	5.5767	6.1626	6.2479	7.4516
		10×10	1.9504	3.5985	5.0782	5.5720	5.9030	7.2281
		20×20	1.9379	3.5493	4.9610	5.4132	5.8064	7.1031
		30×30	1.9356	3.5402	4.9397	5.3838	5.7883	7.0784
	Liew [27]		1.9393	3.5939	4.8755	5.4855	5.7691	7.1177

in table 13.4, while the corresponding convergence study for CCCC rectangular laminate is performed in table 13.5. It can be seen that a faster convergence is obtained for higher t/b ratios irrespective of a/b ratios. In both SSSS and CCCC cases the results converge well to Liew [27] results.

The MATLAB code problem21.m for this case is listed next.

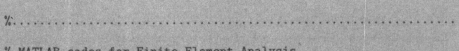

```
%...............................................................

% MATLAB codes for Finite Element Analysis
% problem21.m
% free vibrations of laminated plates
% See reference:
% K. M. Liew, Journal of Sound and Vibration,
% Solving the vibration of thick symmetric laminates
% by Reissner/Mindlin plate theory and the p-Ritz method, Vol.198,
% Number 3, Pages 343-360, 1996
```

Table 13.5 Convergence study of frequency parameters $\bar{\omega} = (\omega b^2/\pi^2)\sqrt{(\rho h/D_0)}$ for three-ply $(0°/90°/0°)$ clamped CCCC rectangular laminates

			Modes					
a/b	t/b	Mesh	1	2	3	4	5	6
1	0.001	5×5 Q4 elements	16.6943	22.0807	51.0268	57.7064	59.9352	76.5002
		10×10	15.1249	18.4938	27.6970	42.6545	44.3895	45.5585
		20×20	14.7776	17.8233	25.2187	37.5788	39.9809	41.6217
		30×30	14.7151	17.7061	24.8194	36.4114	39.5190	41.1433
	Liew [27]		14.6655	17.6138	24.5114	35.5318	39.1572	40.7685
	0.20	5×5 Q4 elements	4.5013	7.3324	7.9268	9.4186	11.9311	12.3170
		10×10	4.4145	6.8373	7.6291	9.2078	10.3964	11.4680
		20×20	4.3931	6.7178	7.5509	9.1264	10.0084	11.1927
		30×30	4.3891	6.6959	7.5363	9.1104	9.9377	11.1415
	Liew [27]		4.4468	6.6419	7.6996	9.1852	9.7378	11.3991
2	0.001	5×5 Q4 elements	5.7995	15.0869	15.1794	20.9280	47.7268	48.0954
		10×10	5.2624	11.3863	11.4494	15.5733	23.0606	23.2217
		20×20	5.1435	10.7292	10.7872	14.6188	20.3457	20.4857
		30×30	5.1221	10.6156	10.6727	14.4537	19.9064	20.0430
	Liew [27]		2.3618	6.6252	6.6845	9.4470	14.2869	16.3846
	0.20	5×5 Q4 elements	3.2165	4.4538	6.4677	6.6996	7.0825	7.8720
		10×10	3.0946	4.2705	5.9857	6.0821	6.7327	7.8602
		20×20	3.0646	4.2207	5.8436	5.9323	6.6221	7.6291
		30×30	3.0591	4.2114	5.8170	5.9049	6.6009	7.5656
	Liew [27]		3.0452	4.2481	5.7916	5.9042	6.5347	7.6885

```
% antonio ferreira 2008

% clear memory
clear all;colordef white;clf

% materials
thickness=0.001;h=thickness;kapa=pi*pi/12;
rho=1;I=thickness^3/12;
% symbolic computation
syms phi pi

% liew material
e2=1;e1=40*e2;g23=0.5*e2;g13=0.6*e2;g12=g13;
miu12=0.25;miu21=miu12*e2/e1;factor=1-miu12*miu21;

% angles for laminate
```

```
alfas=[0,pi/2,0];% 3 layers
% upper and lower coordinates
z(1)=-(h/2);z(2)=-(h/2)+h/3;z(3)=-z(2);z(4)=-z(1);

% [Q] in 0° orientation
qbarra(1,1,1)=e1/factor;
qbarra(1,2,1)=miu21*e1/factor;
qbarra(2,1,1)=miu12*e2/factor;
qbarra(2,2,1)=e2/factor;
qbarra(3,3,1)=g12;
qbarra(4,4,1)=kapa*g23;
qbarra(5,5,1)=kapa*g13;

% transformation matrix
T=[cos(phi)^2,sin(phi)^2,-sin(2*phi),0,0;...
   sin(phi)^2,cos(phi)^2,sin(2*phi),0,0;...
   sin(phi)*cos(phi),-sin(phi)*cos(phi),cos(phi)^2-sin
     (phi)^2,0,0;...
   0,0,0,cos(phi),sin(phi);...
   0,0,0,-sin(phi),cos(phi)];

% [Q] in structural axes
qBarra=T*qbarra*T.';

for s=1:size(alfas,2)
    for i=1:5
        for j=1:5
            QQbarra(i,j,s)=subs(qBarra(i,j,1),phi,alfas(s));
        end
    end
    Qbarra=double(QQbarra);
end
Q=Qbarra;

%----------------------------------------------------
Astiff(5,5)=0;Bstiff(5,5)=0;Fstiff(5,5)=0;Istiff(5,5)=0;
for k=1:size(alfas,2)
        for i=1:3
        for j=1:3
        Astiff(i,j)=Astiff(i,j)+Q(i,j,k)*(z(k+1)-z(k));
        Bstiff(i,j)=Bstiff(i,j)+Q(i,j,k)*(z(k+1)^2-z(k)^2)/2;
        Fstiff(i,j)=Fstiff(i,j)+Q(i,j,k)*(z(k+1)^3-z(k)^3)/3;
        end
```

```
          end

                for i=4:5
                for j=4:5
                Istiff(i,j)=Istiff(i,j)+Q(i,j,k)*(z(k+1)-z(k));
                end
                end
end
pi=double(pi); % come back to numeric computation

% constitutive matrices
CMembranaMembrana=Astiff(1:3,1:3);
CMembranaFlexao0=Bstiff(1:3,1:3);
CFlexao0Flexao0=Fstiff(1:3,1:3);
CCorte0Corte0=Istiff(4:5,4:5);

% load
P = -1;

%Mesh generation
L  = 1;
numberElementsX=10;
numberElementsY=10;
numberElements=numberElementsX*numberElementsY;

[nodeCoordinates, elementNodes] = ...
    rectangularMesh(2*L,L,numberElementsX,numberElementsY);
xx=nodeCoordinates(:,1);
yy=nodeCoordinates(:,2);
drawingMesh(nodeCoordinates,elementNodes,'Q4','k-');
axis off
numberNodes=size(xx,1);

% GDof: global number of degrees of freedom
GDof=5*numberNodes;

% stiffness and mass matrices
stiffness=formStiffnessMatrixMindlinQ45laminated5dof...
   (GDof,numberElements,...
   elementNodes,numberNodes,nodeCoordinates,CMembranaMembrana,...
CMembranaFlexao0,CFlexao0Flexao0,CCorte0Corte0);

[mass]=...
```

```
        formMassMatrixMindlinQ4laminated5dof(GDof,numberElements,...
        elementNodes,numberNodes,nodeCoordinates,rho,thickness,I);

% boundary conditions
[prescribedDof,activeDof,fixedNodeW]=...
    EssentialBC5dof('cccc',GDof,xx,yy,nodeCoordinates,
        numberNodes);

% eigenproblem: free vibrations
numberOfModes=12;

[V,D] = eig(stiffness(activeDof,activeDof),...
    mass(activeDof,activeDof));
% Liew, p-Ritz
    D0=e2*h^3/12/(1-miu12*miu21);
    D = diag(sqrt(D)*L*L/pi/pi*sqrt(rho*h/D0));
    [D,ii] = sort(D); ii = ii(1:numberOfModes);
    VV = V(:,ii);
    activeDofW=setdiff([1:numberNodes]',[fixedNodeW]);
    NNN=size(activeDofW);

    VVV(1:numberNodes,1:12)=0;
    for i=1:numberOfModes
        VVV(activeDofW,i)=VV(1:NNN,i);
    end

NN=numberNodes;N=sqrt(NN);
x=linspace(-L,L,numberElementsX+1);
y=linspace(-L,L,numberElementsY+1);

% drawing Eigenmodes
drawEigenmodes2D(x,y,VVV,NN,N,D)
```

References

1. R. Courant. Variational methods for the solution of problems of equilibrium and vibration. *Bulletin of the American Mathematical Soceity*, 49:1–23, 1943.
2. J. H. Argyris. Matrix displacement analysis of anisotropic shells by triangular elements. *Journal of the Royal Aeronautical Soceity*, 69:801–805, 1965.
3. R. W. Clough. The finite element method in plane stress analysis. *Proceedings of the 2nd A.S.C.E. Conference in Electronic Computation*, Pittsburgh, PA, 1960.
4. J. N. Reddy. *An introduction to the finite element method.* McGraw-Hill, New York, 1993.
5. E. Onate. *Calculo de estruturas por el metodo de elementos finitos.* CIMNE, Barcelona, 1995.
6. O. C. Zienkiewicz. *The finite element method.* McGraw-Hill, New York, 1991.
7. T. J. R. Hughes. *The finite element method-Linear static and dynamic finite element analysis.* Dover Publications, New York, 2000.
8. E. Hinton. *Numerical methods and software for dynamic analysis of plates and shells.* Pineridge Press, Swansea, 1988.
9. Y. W. Kwon and H. Bang. *Finite element method using MATLAB.* CRC Press, Boca Raton, FL, 1996.
10. P. I. Kattan. *MATLAB Guide to finite elements, an interactive approach.* Springer, Berlin, 2nd ed., 2007.
11. D. L. Logan. *A first course in the finite element method.* Brooks/Cole, USA, 2002.
12. K. J. Bathe. *Finite element procedures in engineering analysis.* Prentice-Hall, Englewood Cliffs, NJ, 1982.
13. R. D. Cook, D. S. Malkus, M. E. Plesha, and R. J. Witt. *Concepts and applications of finite element analysis.* Wiley, New York, 2002.
14. C. M. Wang, J. N. Reddy, and K. H. Lee. *Shear deformable beams and plates.* Elsevier, Oxford, 2000.
15. M. Petyt. *Introduction to finite element vibration analysis.* Cambridge University Press, Cambridge, 1990.
16. J. Lee and W. W. Schultz. Eigenvalue analysis of timoshenko beams and axisymmetric mindlin plates by the pseudospectral method. *Journal of Sound and Vibration*, 269(3–4):609–621, 2004.
17. Z. P. Bazant and L. Cedolin. *Stability of structures.* Oxford University Press, New York, 1991.
18. J. N. Reddy. *Mechanics of laminated composite plates.* CRC Press, New York, 1997.
19. D. J. Dawe and O. L. Roufaeil. Rayleigh-ritz vibration analysis of mindlin plates. *Journal of Sound and Vibration*, 69(3):345–359, 1980.
20. K. M. Liew, J. Wang, T. Y. Ng, and M. J. Tan. Free vibration and buckling analyses of shear-deformable plates based on fsdt meshfree method. *Journal of Sound and Vibration*, 276:997–1017, 2004.

21. J. A. Figueiras. *Ultimate load analysis of anisotropic and reinforced concrete plates and shells.* University of Wales, Swansea, 1983.

22. S. Srinivas. A refined analysis of composite laminates. *Journal of Sound and Vibration*, 30:495–507, 1973.

23. B. N. Pandya and T. Kant. Higher-order shear deformable theories for flexure of sandwich plates-finite element evaluations. *International Journal of Solids and Structures*, 24:419–451, 1988.

24. A. J. M. Ferreira. *Analysis of composite plates and shells by degenerated shell elements.* FEUP, Porto, Portugal, 1997.

25. A. J. M. Ferreira. A formulation of the multiquadric radial basis function method for the analysis of laminated composite plates. *Composite Structures*, 59:385–392, 2003.

26. A. J. M. Ferreira, C. M. C. Roque, and P. A. L. S. Martins. Analysis of composite plates using higher-order shear deformation theory and a finite point formulation based on the multiquadric radial basis function method. *Composites: Part B*, 34:627–636, 2003.

27. K. M. Liew. Solving the vibration of thick symmetric laminates by reissner/mindlin plate theory and the p-ritz method. *Journal of Sound and Vibration*, 198(3):343–360, 1996.

Index